Moeketsi Mpholo · Dirk Steuerwald
Tonny Kukeera
Editors

Africa-EU Renewable Energy Research and Innovation Symposium 2018 (RERIS 2018)

23–26 January 2018, National University of Lesotho On occasion of NULISTICE 2018

Editors
Moeketsi Mpholo
Energy Research Centre
National University of Lesotho
Roma, Lesotho

Tonny Kukeera
Energy Research Centre
National University of Lesotho
Roma, Lesotho

Dirk Steuerwald
Swiss Academy for Development
Biel, Switzerland

https://doi.org/10.1007/978-3-319-93438-9

Preface

Background

With more than 600 million Africans living without access to electricity and over 730 million relying on dangerous, inefficient forms of cooking, fast but sustainable development of African renewable energy markets is needed. Applied research and innovative business roll-outs play a key role in boosting this development, helping to secure energy supply in both rural and urban areas, increasing generation capacities and energy system resilience. Aiming to promote research collaboration, public and private sector participation in applied research and business spin-offs in this field, the Africa-EU Renewable Energy Cooperation Programme (RECP) and the National University of Lesotho (NUL) organised the second edition of the Africa-EU Renewable Energy Research and Innovation Symposium (RERIS 2018). The event took place on occasion of the NUL International Science and Technology Innovation Conference and Expo (NULISTICE) from 23–26 January 2018, attended by 242 participants from more than 30 countries.

Scientific and Organising Committees

RERIS Conference Chair

Dr. Moeketsi Mpholo, Energy Research Centre (ERC), National University of Lesotho, Lesotho

RERIS Scientific Committee

Prof. Benmansour Abdelhalim, Université Aboubekr Belkaid de Tlemcen, Algeria
Prof. Venkata Ramayya Ancha, Jimma University, Ethiopia
Dr. Ed Brown, Loughborough University, UK
Dr. Ben Campbell, Durham University, UK

Prof. Thameur Chaibi, National Research Institute for Rural Engineering, Water and Forestry, Tunisia
Prof. Seladji Chakib, Université Aboubekr Belkaid de Tlemcen, Algeria
Dr. Zivaye Chiguvare, Namibia University of Science and Technology
Dr. Jon Cloke, Loughborough University, UK
Prof. Emanuela Colombo, Politecnico di Milano, Italy
Prof. Abdellah Khellaf, Centre de Développement des Energies Renouvelables, Algeria
Prof. Izael Pereira Da Silva, Strathmore University, Kenya
Prof. Daniel Egbe, Johannes Kepler University Linz, Austria
Prof. Thomas Hamacher, Technische Universität München, Germany
Dr. Jarkko Levänen, Aalto University School of Business, Finland
Prof. Angeles López Agüera, UNESCO-CLRLA/Universidade de Santiago de Compostela, Spain
Prof. Martin Meyer-Renschhausen, Hochschule Darmstadt, Germany
Prof. Joseph Mutale, University of Manchester, UK
Prof. Wikus van Niekerk, Stellenbosch University, South Africa
Prof. Armando C. Oliveira, University of Porto, Portugal
Prof. Marco Rupprich, MCI—The Entrepreneurial School®, Austria
Prof. Abdelilah Slaoui, University of Strasbourg, France
Prof. Anthony Staak, Cape Peninsula University of Technology (CPUT), South Africa
Dr. Dirk Steuerwald, Swiss Academy for Development, Switzerland
Dr. Sandor Szabo, European Commission Joint Research Centre, Italy
Dr. Leboli Zak Thamae, Energy Research Centre, National University of Lesotho
Dr. Daniel Yamegueu, International Institute for Water and Environmental Engineering (2iE), Burkina Faso
Prof. Abdellatif Zerga, Pan African University Institute of Water and Energy Sciences (including Climate Change) (PAUWES), Algeria

Organising Committee

Dr. Thimothy Thamae, National University of Lesotho/NULISTICE lead organiser
Mr. Niklas Hayek, Africa-EU Renewable Energy Cooperation Programme (RECP)/RERIS lead organiser
Dr. Moeketsi Mpholo, National University of Lesotho
Prof. Himanshu Narayan, National University of Lesotho
Mrs. Maleshoane Ramoholi, National University of Lesotho
Dr. Pulane Nkhabutlane, National University of Lesotho
Dr. Mosotho George, National University of Lesotho
Mr. Rets'elisitsoe Thamae, National University of Lesotho
Dr. Puleng Ranthimo, National University of Lesotho
Mr. Kebitsamang Mothibe, National University of Lesotho
Dr. Sissay Mekbib, National University of Lesotho
Mr. Lefa Thamae, National University of Lesotho

Mr. Joshua Takalimane, National University of Lesotho
Mr. Benjamin Freischlad, Africa-EU Renewable Energy Cooperation Programme (RECP)
Mr. Tonny Kukeera, Pan African University Institute of Water and Energy Sciences (PAUWES), Tlemcen, Algeria

Symposium Summary

The panel discussion on **'How to translate research into practice'** addressed a number of challenges that slow down this translation. While the communication between technology developers with electricity distributors is generally on a good level, the needs of the consumers—especially those in rural communities without access to the grid—are not considered sufficiently. In the specific case of Lesotho, its National Strategic Development Plan (NSDP) formulates the aim to foster research—also to address the marginalised consumers—but its implementation remains a challenge. The session on **EU-Africa joint programmes** addressed funding and cooperation opportunities for both researchers and private project developers. An example discussed was Horizon 2020 which provides a total of EUR 30 billion, of which 5.9 billion are allocated to secure clean and efficient energy.

During the various **thematic sessions and poster sessions**, invited speakers presented their research and innovative projects covering the areas of power generation, distribution and transmission; de-centralised and household energy solutions; energy socio-economics; promotion of energy research, innovation, education and entrepreneurship; and energy resource mapping and planning. The conference was concluded with a gala dinner; His Majesty King Letsie III of Lesotho delivered the keynote address. At the gala dinner, NUL officially launched its Energy Research Center (ERC) and the Master of Science in Sustainable Energy programme—both established with the support from the RECP.

Paper Review

A total of 150 abstracts were received from 33 countries (22 from Africa, 10 from EU and 1 from America) towards 30 oral presentation slots. For each abstract, a minimum of two reviews were given by the scientific committee members. The final selection was such that women were given priority and half the papers were from the African delegates while the other half were from the EU delegates.

The presentations' full papers for publication were further subjected to a minimum of two reviews upon which only the papers which the reviewers and editors felt met the expected standard were accepted for this publication.

Eschborn, Germany

Niklas Hayek
Deutsche Gesellschaft für Internationale
Gesellschaft (GIZ) GmbH

Roma, Lesotho

Moeketsi Mpholo
Energy Research Centre (ERC)
National University of Lesotho (NUL)

Acknowledgements

The organisers would like to thank all the conference participants, presenters, scientific committee members and the following sponsors for making RERIS 2018 such a big success:

Contents

Chapter 1
A Thermo-Economic Model for Aiding Solar Collector Choice and Optimal Sizing for a Solar Water Heating System

Tawanda Hove

Abstract The choice of solar collector type to employ and the number of chosen collectors to subsequently deploy, are important planning decisions, which can greatly influence the economic attractiveness of solar water heating systems. In this paper, a thermo-economic model is developed for the computation of a suitable metric that can aid in choosing the most cost-effective collector to use in a solar water heating system and to determine the optimal sizing of the solar water heater components once the choice collector has been picked. The energy-per-dollar comparison metric, calculated as the annual heat energy output of the collector in an average year, at the so-called "sweet-spot" size of the collector array, divided by the annualized life-cycle cost, based on warranty life and collector initial cost, was recommended as instructive for comparing cost-effectiveness of different solar collectors. For the determination of the sweet-spot size of collector to use in a particular solar water heating system, at which the energy-per-dollar is calculated, the Net Present Value of Solar Savings was used as the objective function to maximize. Ten (10) different models of liquid solar thermal collectors (5 flat plate and 5 evacuated tube type), which are rated by the Solar Ratings & Certification Corporation (SRCC), were ranked according to the energy-per-dollar criterion through the thermo-economic model described in this study. At the sweet-spot collector area for the solar water heating system, the corresponding volume of hot water storage tank and the optimal solar fraction are also simultaneously determined. The required hot water storage volume decreases as the deployed collector area increases while the solar fraction increases, with diminishing marginal increase, until it saturates at a value of unity. For the present case study where the required load temperature is 50 °C and the solar water heating system is located in central Zimbabwe (latitude 19° S and longitude 30° E), the selected collector model happened to be a flat-plate type, which achieved the highest energy-per-dollar score

---

OCRID Researcher ID: E 4889-2018

T. Hove (✉)
Department of Mechanical Engineering, University of Zimbabwe, Harare, Zimbabwe
e-mail: tawandahv2@yahoo.co.uk

M. Mpholo et al. (eds.), *Africa-EU Renewable Energy Research and Innovation Symposium 2018 (RERIS 2018)*, Springer Proceedings in Energy,
https://doi.org/10.1007/978-3-319-93438-9_1

of 26.1 kWh/\$. The optimal size of this collector model to deploy in the solar water heating system at the case-study location is 18 m^2 per m^3 of daily hot water demand; with a hot water storage volume of 900 l/m^3; at an optimal solar fraction of 91%. Although the method of this paper was applied only for a solar water heating application of specified operating temperature, at a specified location, it can be applied equally well for any other solar water heating application and at any other location.

Keywords Energy-per-dollar · Thermo-economic model · Collector choice Optimal sizing · Diminishing marginal returns · SRCC-rated

Nomenclature

Δt	Finite time-step period over which the solar process is simulated [s]
T_s	Temperature of hot water storage tank contents at the beginning of the period Δt [°C]
ΔT_s	Temperature gain/loss for storage tank during time-step period Δt [°C]
T_s^+	Temperature of storage tank contents at the end of Δt [°C]
T_a	Ambient temperature [°C]
T_{mains}	Incoming cold water temperature [°C]
T_{load}	Temperature required by the load [°C]
$\dot{m}_s$	Mass rate of water abstraction [kg/s]
$\dot{m}_{load}$	Mass rate of water abstraction required by the load at temperature T_{load} [kg/s]
M	Mass of hot water storage tank [kg]
V_s	Volume of hot water storage tank [m^3]
U_s	Storage tank heat loss coefficient [W/m^2/°C]
As	Storage tank surface area [m^2]
Q_u	Rate of useful heat generated by the solar energy collector
$\dot{L}_s$	Rate of removal of solar-generated heat [Watts]
G_T	Global solar irradiance incident on the plane of the collector [W/m^2]
A_C	Solar collector gross area [m^2]
F_R	Collector heat removal factor [–]
$(\tau\alpha)_n$	Transmittance-absorption product [–]
U_L	Collector heat loss coefficient [W/m^2/°C]
$K_{\tau\alpha}$	Angle of incidence modifier [–]
S_F	Solar fraction; the fraction of required heat contributed by solar energy [–]
C_C	Collector cost per unit area [\$/m^2]
C_S	Storage tank cost per unit volume [\$/m^3]
i	Annual interest rate [–]
j	Annual inflation rate [–]
d	Annual discount rate [–]
Q_{annual}	Energy generated by solar collector over an average year [kWh/m^2/yr]
C_{annual}	Annualized cost of collector based on warranty period [\$/m^2/yr]

1.1 Introduction

Worldwide, the heating of water to low and medium temperatures using solar thermal energy has gained popularity for many residential, commercial and industrial applications. This is because of the numerous favourable characteristics of solar water heating, which result in a large displacement of conventional energy sources in an economically and environmentally sustainable way. Solar thermal water heaters are a prudent option where, among other considerations, the cost of conventional energy for heating water is higher than $0.034/kWh; daily average solar irradiation is higher than 4.5 kWh/m^2 and where energy security is important (e.g. where there is interruptible supply of conventional energy) [1]. In Zimbabwe, the price of electricity is $0.11/kWh for domestic customers consuming 50–300 kWh/month and from $0.04/kWh (off-peak) to $0.13/kWh (peak period), for time-of-use customers [2]. Solar radiation is abundant (an annual average of 5.6–7 kWh/day) and ubiquitously distributed over the country [3]. Electricity, which is the conventional energy for heating water, is in short supply, with frequent load-shedding [4] events and reliance on importation of the commodity. With this scenario, it is not surprising that the Zimbabwe Government has recently launched the National Solar Water Heater Program [5], where water heating by electric geysers will be substituted by solar thermal water heating systems. This policy initiative is perceived to have many favourable outcomes including the reduction in national electricity consumption, improvement of consumer economics as well as a significant contribution to mitigating environmental degradation. A challenge that exists, however, is the high initial costs associated with owning a solar water heating system, which calls for cost-efficient selection of solar water heating components and their sizing, in order to maximise life-cycle economic benefits.

A solar water heating system essentially consists of a solar thermal collector and a water storage tank although other balance-of-system components such circulation pump, pipe-work and control system may also variably be included. Of the two main components, the solar collector, which is at the heart of the solar water heating system, has greater influence on system performance and costs up to twice or more times the cost of the storage tank [6]. Correct selection of the type and size of the collector to employ in a solar water heating system has great influence in the economic viability of the solar water heating system, as this determines the trade-off between system cost and solar fraction (conventional energy costs displaced). In addition the temperature achieved by the solar water heating system, which is determined by the type and size of the solar collector employed, have some influence on the size of the storage tank required, as a large collector area (higher operating temperature) should result in less required volume of storage tank [7]. Hence it is important to carefully choose the type and optimal size of a solar collector to use for a particular solar water heating application and climatic conditions.

Different attributes may be exclusively or jointly used to appraise the prudency of the choice of the solar thermal collector to purchase for a given application. These attributes include initial cost; energy performance; warranty (which is some guarantee for longevity) and others such as chance of overheating, ability to shed snow and wind-load structural capability. However, a more instructive metric to use when selecting between different types of collectors is the *energy-per-dollar* metric [8], as it can be made to combine an important-few of collector attributes, which determine the collector's life-cycle cost-effectiveness, into one metric. In this study, a thermo-economic model is developed to guide the selection of the brand of solar collector(s) to be used in a solar water heating application, for a given required hot water temperature under given climatic conditions. The model is also used to determine the optimal size of collector area to deploy in the solar water heating system, together with corresponding volume of hot water storage tank and solar fraction at optimal collector size. A modified approach to defining and calculating the energy-per-dollar metric, which compares annual energy output of the collector under certain operating temperatures and annualized collector costs, taking into account the assured operating longevity of the collector (warranty life), is used in this study for ranking the cost-effectiveness of different collectors. The Net Present Value of Solar Savings (NPVSS) is used as the objective function to maximize for the selection of the optimal size of collector area to be deployed in the solar water heating system.

1.2 Solar Water Heating System Thermal Model

A solar water heating system can be represented by the schematic on Fig. 1.1. At the heart of a solar thermal system is the solar collector. It absorbs solar radiation, converts it into heat, and transfers useful heat to a well-insulated hot water storage tank. A pump is sometimes needed to circulate the heat around the system, but in

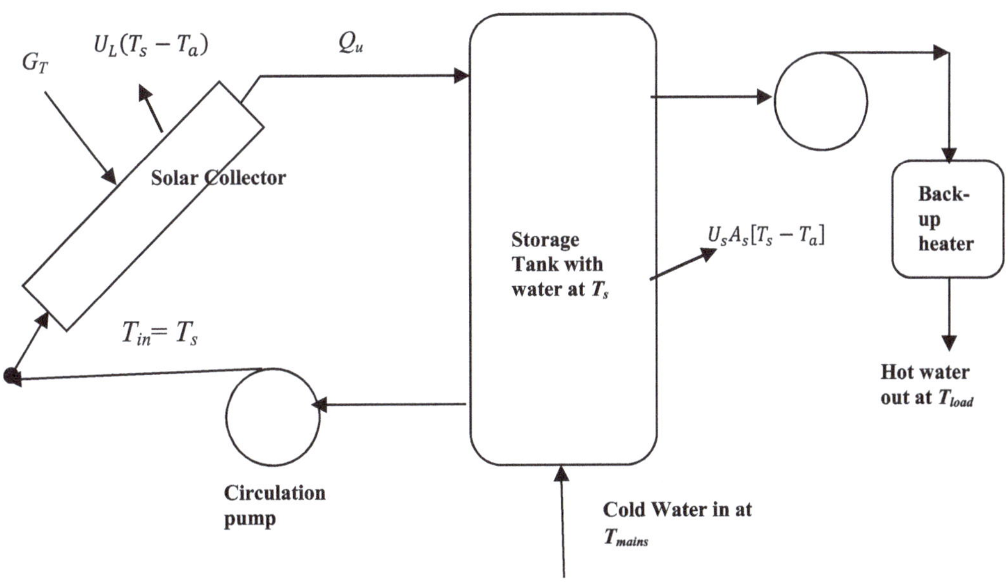

Fig. 1.1 Schematic of a solar water heating system

thermo-siphon systems this is not required as natural gravity circulation is used. In the present model, a secondary smaller storage tank receives water heated by the solar loop and passes on to the load after, if required, boosting the water temperature to the required load temperature, T_{load}.

For the system on Fig. 1.1, assuming a well-mixed storage tank of mass M and fluid specific heat capacity C_P, a simple energy balance can be expressed by the following differential equation:

$$(MC_P)_s \frac{dT_s}{dt} = Q_u - \dot{L}_s - U_s A_s [T_s - T_a] \tag{1.1}$$

Equation (1.1) states that the rate of change in the internal energy of the storage tank is equal to the energy interactions taking place over a time step.

The energy interactions are solar-collector-generated heat input Q_u; the rate of heat removal ($\dot{L}_s$) and the storage tank heat losses $(U_s A_s [T_s - T_a])$. One can use simple Euler integration [9], (i.e., rewriting the temperature derivative as $\frac{T_s^+ - T_s}{\Delta t}$ and solving for the tank temperature at the end of a time increment). That is:

$$T_s^+ = T_s + \frac{\Delta t}{(MC_P)_s} \left(Q_u - \dot{L}_s - U_s A_s [T_s - T_a] \right) \tag{1.2}$$

A time step $\Delta t = 1\,\text{h}$ is conveniently used since one hour is the smallest time resolution of solar radiation data commonly available.

In Eq. (1.1), solar input Q_u is determined using Eq. (1.3). For a collector of area A_C with solar irradiance G_T incident on its plane, the rate of useful thermal energy gained is given by the well-known Hottel-Whillier-Bliss equation:

$$Q_u = A_c \left\{ G_T K_{\tau\alpha} F_R (\tau\alpha)_n - F_R U_L (T_s - T_a) \right\} \tag{1.3}$$

In Eq. (1.3), $(\tau\alpha)_n$ is the normal-incidence transmittance-absorptance product of the collector—the fraction of incoming radiation absorbed at normal incidence. The ratio of absorbed incident radiation at any angle of incidence θ, to that at normal incidence is called the '*incident angle modifier*', $K_{\tau\alpha}$. The collector heat loss coefficient U_L, is a measure of the rate of collector heat loss per unit area and per unit temperature difference between the collector and the ambient. If the reference collector temperature used is the fluid inlet temperature, the collector heat removal factor F_R is applied as a multiplier to convert the energy gain if the whole collector were at fluid inlet temperature to the actual collector energy gain. The collector fluid inlet temperature is assumed to be equal to the average storage tank temperature T_S and the heat loss is to the ambient environment at temperature T_a.

Data for the collector characteristics $F_R(\tau\alpha)_n$ (the y-intercept of the collector efficiency curve); $F_R U_L$ (the negative of the slope of the efficiency curve) and the incident-angle-modifier $K_{\tau\alpha}$ are indispensable for the prediction of collector energy performance under specific operating-temperature and climatic conditions. These data are sometimes provided on manufacturers' product data sheets. However, the authentic information should be obtained from internationally accredited certifying bodies such as the Solar Rating & Certification Corporation (SRCC) [10]. The SRCC clearly gives values of $F_R(\tau\alpha)_n$ and $F_R U_L$ with respect to the gross area for each tested collector together with the variation of $K_{\tau\alpha}$ with angle of incidence of beam radiation on the collector. Figure 1.2a, b compare typical plots of collector efficiency curves and incidence angle of modifiers typical for flat plate and evacuated tube collectors. The plots are based on SRCC data for collectors from the same manufacturer. With reference to gross area, flat plate collectors generally have much greater values of both $F_R(\tau\alpha)_n$ and $F_R U_L$, making them superior to evacuated tube collectors for relatively low operating $\frac{(T_s - T_a)}{G_T}$ (e.g. for domestic solar water heating applications in warm climates with high solar radiation). On the other hand, the optical efficiency of flat plate collectors is affected more adversely by increase in angle of incidence than evacuated tube collectors. It is always interesting to find out

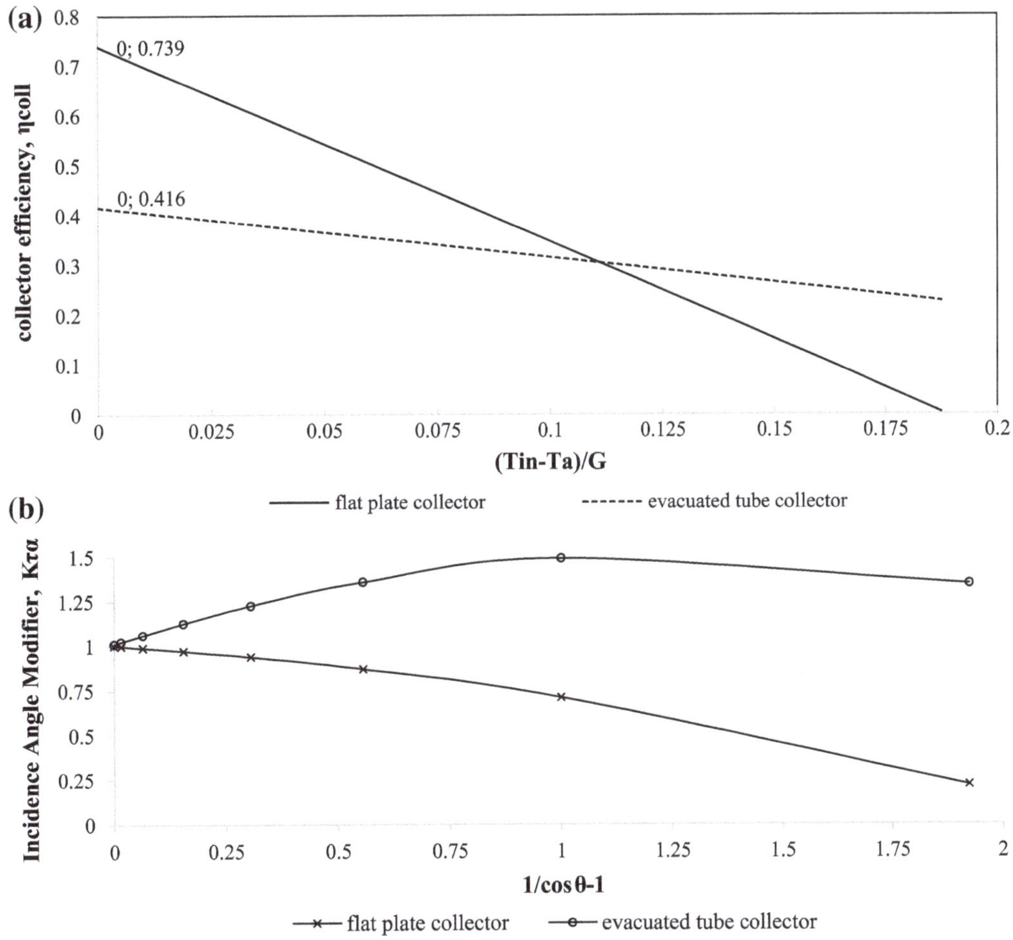

Fig. 1.2 **a** Typical efficiency curves with respect to gross area for flat plate and evacuated tube collectors. **b** Typical variation of $K_{\tau\alpha}$ with 1/cosθ−1 for flat plate and evacuated tube collectors. At angle of incidence $\theta = 0°$, the incidence angle modifier $K_{\tau\alpha} = 1$ and at $\theta = 90°$, $K_{\tau\alpha} = 0$

which of the characteristics has greater influence on energy output and which collector "wins" for specific operating and climatic conditions.

Continuing with expounding Eq. (1.1), the rate of heat extraction from the solar part of the system L_s is considered next. The rate of heat removal from the solar-side storage tank is given by:

$$\dot{L}_s = \dot{m}_s C_P (T_s - T_{mains}) \tag{1.4}$$

In Eq. (1.4), $\dot{m}_s$ (kg/s) is the mass rate of water withdrawal; C_P is the specific heat capacity of water; T_s is the temperature of water in the storage tank and T_{mains} is temperature of the cold water coming from the mains supply. The mass rate of water withdrawal $\dot{m}_s$ depends on the amount of water withdrawal $\dot{m}_{load}$ required by the system user at load temperature T_{load} and the comparison between T_s and T_{load}. If $T_s > T_{load}$, then the required water withdrawal $\dot{m}_s < \dot{m}_{load}$(less water needs to be withdrawn from the system and the volume is made up by cold water from outside the system). On the other hand, if $T_s < T_{load}$ an amount of hot water $\dot{m}_{load}$ needs to be withdrawn and its temperature is boosted by the auxiliary heater. Therefore, one can be write:

$$\dot{m}_s = MIN\left\{\frac{T_{load} - T_{mains}}{T_s - T_{mains}}; 1\right\} \times \dot{m}_{load} \tag{1.5}$$

The solar fraction is the ratio of the amount of input heat energy contributed by the solar energy system to the total input energy required for the water heating application for a specified period of time. Then instantaneous solar fraction S_F can be written:

$$S_F = \frac{\dot{m}_s (T_s - T_{mains})}{\dot{m}_{load}(T_{load} - T_{mains})} \tag{1.6}$$

The hourly conventional energy displaced by solar energy each hour is $3600 \times S_F \times \dot{m}_{load}(T_{load} - T_{mains})$, if a constant rate of hot water withdrawal is assumed during the hourly period. The average monthly contribution of displaced energy is obtained by summing the hourly contributions over the average solar radiation day of the month, Klein [11], and multiplying this sum by the number of days of the month. The annual solar contribution of required thermal energy L_{annual} is obtained by summing the monthly contributions.

1.3 Economic Model

In this section, the energy-per-dollar of a solar collector and the Net Present Value of Solar Savings are defined.

In an economy where the prevailing interest rate is i per annum and the rate of inflation is j per annum, it can be shown that the discount rate d is related to i by the expression:

$$d = \frac{1 + i}{1 + j} - 1 \tag{1.7}$$

The annualized cost of a solar collector array of area A_c (m^2) and cost per unit area C_c (\$/m^2), with annual operation and maintenance cost OM over w warranted operating years, is then given by:

$$C_{annual} = A_c C_c \frac{d}{1 - (1 + d)^{-w}} + OM \tag{1.8}$$

In Eq. (1.8) the initial cost of the collector $A_c C_c$ is multiplied by the cost recovery factor $\frac{d}{1-(1+d)^{-w}}$ and the product added to the annual operation and maintenance cost, in order to obtain the total annualized cost.

The annual thermal yield of the solar collector array Q_{annual} is the instantaneous rate Q_u integrated over the whole year. The marginal heat productivity (thermal yield per unit area) of the collector, when employed in the closed system of the solar water heater, diminishes as the area of collectors deployed in a solar heating system increases [12], as shown on Fig. 1.3. There is a corresponding diminishing marginal increase in solar fraction and, as result, the solar fraction follows the 'elbowed' curve shown on Fig. 1.3. Although engineering intuition suggests that there is a 'sweet spot' for sizing the solar collector array, i.e. around the 'elbow' of the solar fraction curve, there is need to for a way to pin-point the exact coordinates of this optimal point.

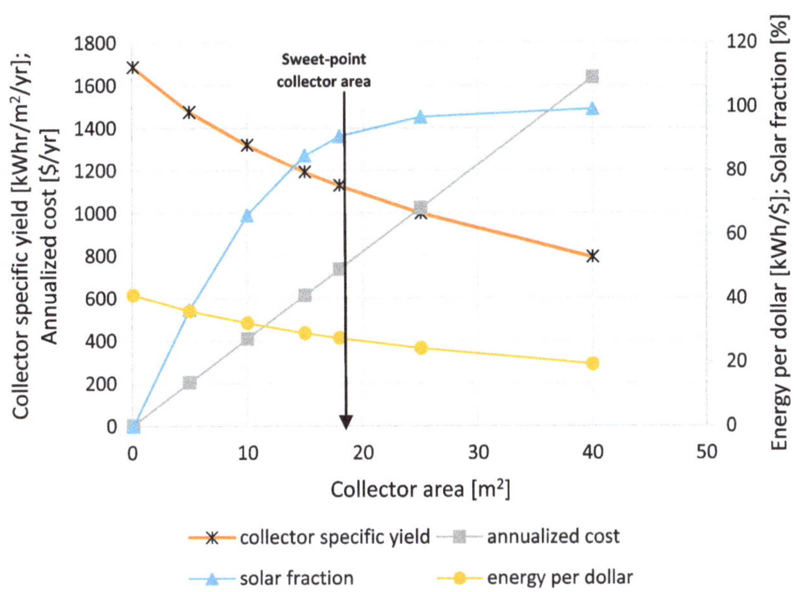

Fig. 1.3 Variation of specific collector heat yield, solar fraction, annualized cost, solar fraction and energy-per-dollar with collector area deployed in a solar water heating system

A suitable metric to compare the level of cost-effectiveness of different collectors is the *energy-per-dollar (epd)*, which divides the average annual yield of the collectors by the annualized life-cycle (warranty-life) cost, Q_{annual}/C_{annual} at the sweet-spot collector area and solar fraction.

$$epd = \frac{Q_{annual}}{C_{annual}} \tag{1.9}$$

When selecting among collectors to use for a specified application (hot water temperature and climatic conditions) the collector model that gives the highest energy-per-dollar, as calculated from Eqs. (1.6, 1.7, 1.8, 1.9), should be preferred. The annualized-cost definition of the energy-per dollar allows for a comparison of the cost-effectiveness of different solar collectors that takes into account for the differences in guaranteed service life (warranty) of the collector. The energy-per-dollar for each candidate collector should be calculated at its sweet spot in the solar water heating system, for objective comparison, as again illustrated on Fig. 1.3.

The precise selection of the sweet spot of solar fraction for a solar water heating system is done by using the Net Present Value of Solar Savings (*NPVSS*) as the objective function to maximize. The NPVSS for a solar water heating system replacing electricity as the heating fuel is given by:

$$NPVSS = \frac{(L_{annual} \times P_E)}{\eta_E} \frac{(1 - [1 + d]^{-n})}{d} - (A_C C_C + V_S C_S + C_{BOS}) \tag{1.10}$$

In Eq. (1.9), the new terms are the price of electricity P_E [$/kWh]; the electric-to-heat efficiency η_E; the hot water tank storage volume V_s and the cost of the storage tank per unit volume C_s. The cost of balance-of-system components C_{BOS} (which also includes installation labour costs) may be taken as a discretionary percentage of the cost $A_C C_C + V_S C_S$. The annual solar energy contributed L_{annual} is expressed in kWh.

To obtain the sweet spot of solar fraction, the collector area is progressively varied, which in turn alters the storage tank temperature T_s and the solar fraction and L_{annual}. The required storage tank volume V_s also varies as dictated by Eq. (1.4), decreasing as the collector area (and tank temperature) increases. Therefore the NPVSS varies with increase in collector area, first increasing as collector area increases, and then decreasing when the collector area is increased beyond its optimal (sweet-spot) value.

1.4 Materials and Methods

A computer spreadsheet program was designed to handle the thermal and economic calculations necessary for appraisal of solar collectors and optimal sizing of the solar water heating system. The computations are done on hourly time-steps for the average day of each month. Starting with only the monthly-average daily global horizontal irradiation (GHI), obtained in this case from the Zimbabwe database developed in [13], the hourly-average GHI for each month was obtained by the model of Collares-Pereira and Rabl [14]. The monthly average diffuse radiation on a horizontal surface was obtained from global radiation by using a diffuse-ratio versus clearness index correlation function suitable for Zimbabwe [15]. The hourly-average diffuse radiation values were determined in a similar fashion as the hourly GHI. Collector-plane hourly irradiation was obtained by applying the isotropic tilted-plane model of [16], with ground albedo fixed at 0.2.

Hourly ambient temperature was generated from monthly-average minimum and maximum ambient temperature using the prediction-model of [17]. The temperature of the cold water entering the solar water heating system T_{mains} was assumed to be equal to the temperature of the soil in which the water mains is buried and predicted through its correlation with air temperature [18].

The required monthly average meteorological data for the case study site (Kwekwe, Zimbabwe) is shown in Table 1.1.

To obtain the surface area of the storage tank (over which heat losses occur) with only knowledge of the tank volume, a cylindrical tank with height twice the diameter is assumed. Then it can be shown that the tank surface area A_s is related to the volume by:

$$A_s = 5.812 V_s^{2/3} \tag{1.11}$$

The diurnal variation of hot water consumption that was used is adapted from the diurnal pattern measured by [19] for hotels in South Africa and is shown on Fig. 1.4. The seasonal variation of daily hot water demand was again adopted from South Africa data given by [20].

The computations for the hourly variation in heat contribution proceeded as shown on Table 1.2. The determination of the storage tank temperature during the first hour of the day is done iteratively, given that the concept of the average day implies that the performance of the solar heating system is identical for all days in a given month. Therefore, the tank temperature in hour 1 is equal to the tank temperature in hour 24 plus the temperature gain/loss in hour 24 (i.e. $T_s(hour1) = T_s(hour24) + \Delta T_s(hour24)$. This also implies that the sum of the hourly temperature gains/losses over the average day is equal to zero (i.e. $\sum_1^{24} \Delta Ts = 0$), Hove and Mhazo [21], as indicated in Table 1.1.

Table 1.1 Monthly average meteorological data for Kwekwe, latitude 19° S and longitude 30° E

Month	January	February	March	April	May	June	July	August	September	October	November	December
Avg. Temperature (°C)	21.9	21.7	20.9	19.8	16.8	14.2	14.2	16.5	20	22.8	22.3	22
Min. Temperature (°C)	16.3	16	14.5	12.6	8.6	6	5.6	7.6	11.2	14.8	16	16.4
Max. Temperature (°C)	26.4	26	25.3	24.6	22.5	20.3	20.3	22.9	26.7	29.1	27.4	26.4
Global horizontal irradiation [kWh/m^2/day]	6.65	6.48	6.35	5.83	5.27	4.90	5.07	6.02	6.83	7.18	6.84	6.35
Diffuse horizontal irradiation [kWh/m^2/day]	2.4	2.3	2.0	1.5	1.2	1.1	1.0	1.2	1.5	2.0	2.4	2.6

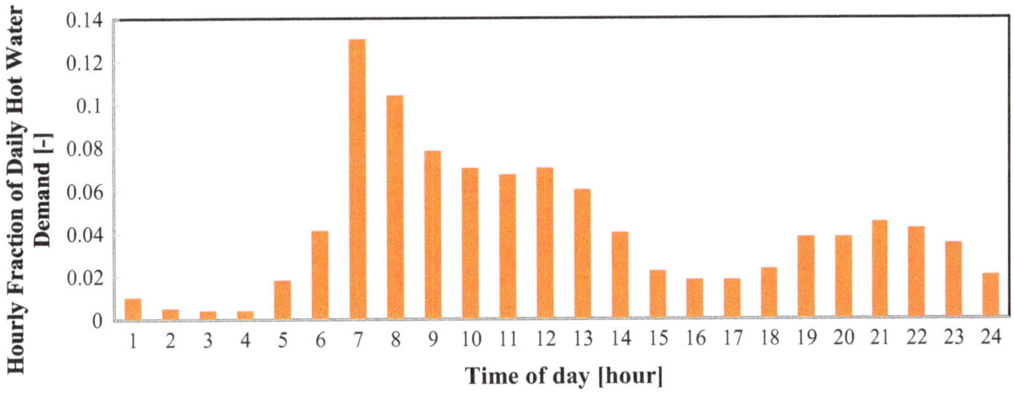

Fig. 1.4 Normalized diurnal variation of hot water demand (at temperature 50 °C) for a hotel in Southern Africa. Adapted from Rankin and Rousseau [19]

1.5 Results and Discussion

Ten (10) different brands of non-concentrating collectors, including both flat plate and evacuated tube collectors, all of which have been rated under the OG-100 collector certification program of the Solar Ratings & Certification Corporation (SRCC), were ranked by the energy-per-dollar metric of Eq. (1.9). In order to make a fair comparison, the energy-per-dollar metric for all the collector arrays was calculated when the collectors are sized such that heat *1000 L of hot water to a temperature 50 °C at an annual solar fraction that results in maximizing the net present value of solar savings*. The reference area used for specifying collector efficiency parameters $F_R\tau\alpha$ and F_RU_L and other area-related parameters in Table 1.3, is the *gross collector area*.

The actual names of the collector manufacturers and brands are withheld, but the individual manufacturers will surely recognize their product from its performance characteristics given on Table 1.3. The collector with rank 1 with an energy-per-dollar score of 26.1 kWh/$ should rationally be selected ahead of its competitors for the water heating system, at the specified temperature (50 °C), for climatic conditions similar to those prevailing in central Zimbabwe. Other selection metrics, which are normally used by different individuals in deciding which collector to buy, are also shown on Fig. 1.3. These metrics include the collector cost/unit area; the collector efficiency parameters ($F_R\tau\alpha$, F_RU_L and the incidence angle modifier function); the warranty period offered by the supplier and the collector footprint area per unit volume of hot water demand. However, it is argued here that the energy per dollar is the most objective metric for selecting among collectors. Using individual collector attributes as selection criteria may result in a non-objective decision about the most cost-effective collector to employ. For example the collector with the lowest cost per gross area ($157/m^2) ranked 10th according to the energy-per-dollar criterion because of its other less favourable attributes such as low optical efficiency and short warrant period. On the other, a collector which is more than twice as expensive ($345/m^2) ranked a respectable 4th,

Table 1.2 Tabulation of procedure for calculating solar water heater system performance

Hour	G_T [W/m^2]	Ta [°C]	Ts [°C]	$K_{\tau\alpha}$ [–]	Q_u [Wh]	$UA_s(T_s - T_a)$ [Wh]	m_{load} [kg]	Load [Wh]	m_s [kg]	ΔT_s [°C]	L_s [Wh]	Solar fraction [–]
0–1	0	20.0	52.4	0.000	0	473	10	401	9	−0.8	401	1.000
1–2	0	17.5	51.6	0.000	0	499	5	200	5	−0.6	200	1.000
2–3	0	14.6	51.0	0.000	0	531	4	160	4	−0.6	160	1.000
3–4	0	11.9	50.4	0.000	0	562	4	160	4	−0.6	160	1.000
4–5	0	9.8	49.8	0.000	0	583	18	721	18	−1.1	716	0.993
5–6	0	8.7	48.7	0.000	0	583	41	1642	41	−1.9	1578	0.961
6–7	66	8.8	46.8	0.522	0	555	130	5208	130	−4.5	4722	0.907
7–8	257	10.0	42.2	0.859	657	471	104	4166	104	−2.6	3226	0.774
8–9	473	12.2	39.6	0.921	3857	401	78	3125	78	1.1	2181	0.698
9–10	682	14.9	40.7	0.920	6522	377	70	2804	70	3.5	2047	0.730
10–11	847	17.8	44.3	0.950	8834	387	67	2684	67	5.4	2235	0.833
11–12	938	20.2	49.6	0.993	10,318	429	70	2804	70	6.1	2772	0.989
12–13	938	21.9	55.7	0.993	10,004	494	60	2404	51	5.8	2404	1.000
13–14	847	22.5	61.5	0.950	7951	570	40	1602	30	4.5	1602	1.000
14–15	682	21.9	66.0	0.920	5231	644	22	881	15	2.8	881	1.000
15–16	473	20.2	68.9	0.921	2364	710	18	721	12	0.5	721	1.000
16–17	257	17.8	69.3	0.859	0	753	18	721	12	−1.6	721	1.000
17–18	66	14.9	67.7	0.522	0	771	23	921	15	−1.9	921	1.000
18–19	0	12.2	65.8	0.000	0	784	38	1522	26	−2.6	1522	1.000
19–20	0	10.0	63.3	0.000	0	778	38	1522	27	−2.5	1522	1.000
20–21	0	8.8	60.8	0.000	0	759	45	1803	34	−2.7	1803	1.000
21–22	0	8.7	58.1	0.000	0	721	42	1682	34	−2.4	1682	1.000
22–23	0	9.8	55.7	0.000	0	669	35	1402	30	−2.0	1402	1.000
23–24	0	11.9	53.7	0.000	0	610	20	801	18	−1.3	801	1.000
Sum/ Average	6524				55,738	14,116	1000	40,059	904	0	36,382	0.91

Table 1.3 Ranking of ten (10) different brands of collectors with respect to energy-per-dollar for a solar water heating application with required hot water temperature of 50 °C, 90% solar fraction and located in central Zimbabwe (latitude 19° S, longitude 30° E)

Rank	Collector type	Cost/area [$/m^2]	$F_R\tau\alpha$	F_RU_L [W/m^2/°C]	K_0	K_1	K_2	Warranty [years]	Annual energy [kWh/m^2]	Annualized cost [$/m2]	Energy/$ [kWh/$m^2]	Required collector area [m^2/m^3]	Solar fraction
1	FPC	220	0.739	3.92	1.001	−0.166	−0.125	10	1128	43.15	26.1	17.0	0.91
2	FPC	303	0.774	3.08	1	0.000	0	10	1399	59.40	23.5	14.0	0.93
3	FPC	242	0.737	4.65	1.002	−0.201	−0.0006	10	1105	47.46	23.3	16.0	0.87
4	FPC	345	0.758	4.14	1.001	−0.287	0.003	10	1198	67.66	17.7	13.5	0.82
5	ETC	175	0.409	1.68	0.999	1.383	−0.992	5	895	51.90	17.2	22.0	0.93
6	FPC	347	0.76	6.22	1.001	−0.035	−0.175	10	1068	68.05	15.7	16.0	0.85
7	ETC	433	0.458	1.58	1	1.313	−1.043	15	1065	71.80	14.8	15.5	0.73
8	ETC	361	0.416	1.08	1.011	0.808	−0.33	10	1025	70.80	14.5	13.0	0.84
9	ETC	211	0.406	1.75	1	1.145	−0.606	5	898	62.58	14.4	22.5	0.94
10	ETC	157	0.383	2.04	1.002	−0.043	0.011	5	658	46.57	14.1	26.0	0.86

K_0, K_1 and K_2 are coefficients of the SRCC-data derived angle of incidence modifier function: $K_{\tau\alpha} = K_0 + K_1(1/\cos\theta - 1) + K_2(1/\cos\theta - 1)^2$
FPC flat plate collector, *ETC* evacuated tube collector

because of its other more favourable attributes such as higher optical efficiency and longer guaranteed longevity (warrant period).

Table 1.3 shows that flat plate collectors (FPC) generally rank higher than evacuated tube collectors (ETC) for solar water heating applications for the climatic conditions considered, according to the energy-per-dollar criterion. For a comparable solar energy output (comparable solar fraction), e.g. collector 2 and collector 5, the required collector footprint area for an FPC is considerably smaller than that for an ETC because of the higher gross area efficiency of the FPC in the considered temperature-gain range. The required footprint area may be used as a secondary selection criterion where collector-placement space is an issue.

Once the most cost-effective collector to buy for a specific application and climatic conditions has been selected, the next questions to be answered are; which the most economic area is to deploy for the solar water heating system; what size of storage tank should fit this collector area and what is the optimal solar fraction. These questions can be answered by employing a thermo-economic model, like the one used in this paper, and making the net present value (NPV) of solar savings the objective function.

Figure 1.5 shows the variation solar fraction, NPV and storage-ratio (volume of storage tank to daily hot water demand at load temperature), applied for a solar water heating system at the selected location and employing the top ranked collector on Table 1.3. The value of the collector area for which the net present value is maximum is the optimal collector area to be specified for the solar water heating system. The values of the normalized storage tank volume and the solar fraction read from the chart corresponding to the optimal collector area, at this "sweet spot", are the optimal specifications for these variables. For instance, at the selected location in central Zimbabwe (latitude 19° South and longitude 30° East), when using the top-ranked flat plate collector on Table 1.3 to heat water to 50 °C, the optimal collector area is 18 m^2 per every m^3 of daily hot water consumed.

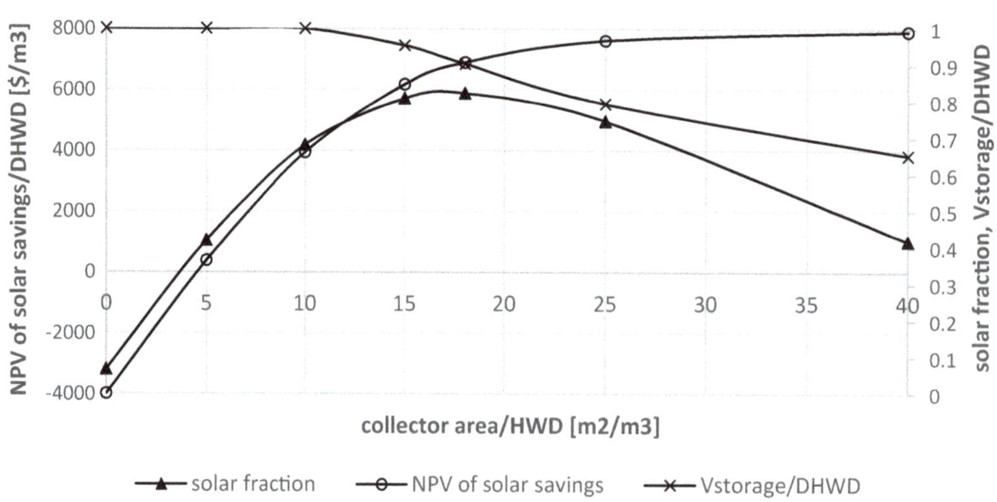

Fig. 1.5 Determination of optimal values of collector area; required storage volume and solar fraction for a solar water heating system

The required storage volume is 900 L (0.9 × 1000 L) per m^3 of daily hot water demanded. The required hot water storage volume decreases as the deployed collector area increases, while the solar fraction increases, with diminishing marginal returns, until it saturates at a value of unity. At optimal design point, the solar fraction is 0.91 and the amount of solar savings in present value terms is $5881 for every m^3 of daily hot water consumed. Due to diminishing marginal thermal energy returns, attempting to increase the solar fraction by only 8%, from the optimal 91 to 99%, will require increasing the collector area from 18 to 40 m^2, resulting in the net present value diminishing from $5881 to $1019. On the other hand, reducing the size of the collector area below 4 m^2, will result in a loss-making solar water heating system—the net present value will be less than zero.

The input data used to come up with the results of Fig. 1.5 is listed on Table 1.4.

The diurnal temperature performance of the 1000-L solar water heating system is shown on Fig. 1.6, together with the simulated incident solar irradiance and ambient temperature. For this particular optimally sized solar water heating system, the temperature of water in the storage tank reaches a maximum of 69 °C (19 °C above the load temperature) and a minimum of 40 °C (10 °C below the load temperature). With the assumed hot water withdrawal pattern (Fig. 1.4), the back-up heater will be required between 6 am and 12 pm, when the temperature of the water supplied by the solar system is less than load temperature 50 °C. The storage tank temperature dips to a low at about 9 am due to large hot water

Table 1.4 Values of input parameters

Parameter	Value	Unit	Parameter	Value	Unit
Average month	May	[–]	Tank heat loss	2.5	W/m²/°C
Latitude	−19	Degrees	Electric-heat efficiency	75	%
Collector tilt	24	Degrees	Collector cost	220	$/m²
Horizontal irradiation	5.3	kWh/m²/day	Tank cost	2000	$/m³
Diffuse irradiation	1.2	kWh/m²/day	Installation/Capital cost	35	%
Maximum monthly T_a	22.5	°C	Maintenance/Capital cost	1	%
Minimum monthly T_a	8.6	°C	Electricity price	0.11	$/kWh
Daily hot water demand	1000	Litres	Interest rate	13	%/annum
T_{mains}	15	°C	Inflation rate	1	%/annum
T_{load}	50	°C	Collector warranty period	10	Years
Collector $F_R\tau\alpha$	0.739		Economic time horizon	20	Years
Collector F_RU_L	3.9	W/m²°C			

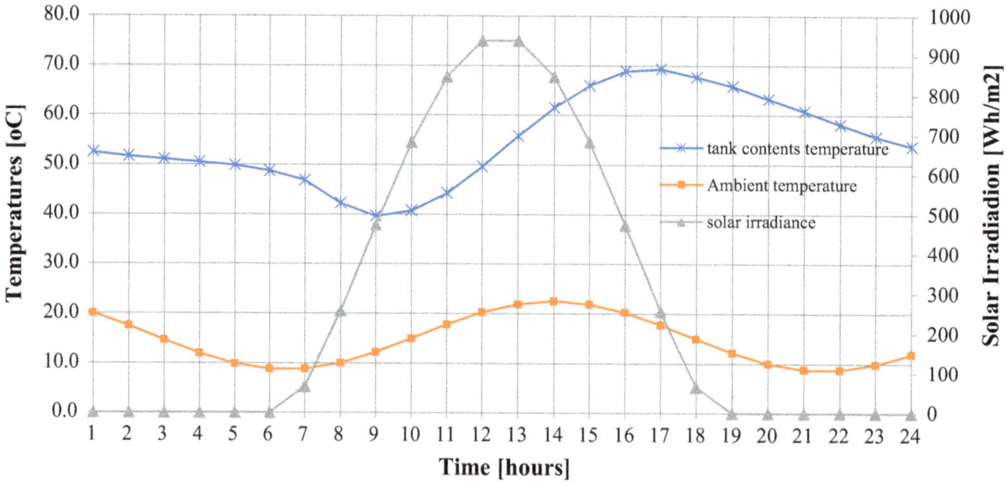

Fig. 1.6 Simulated diurnal variation of storage tank temperature for the optimally sized solar water heating system for the month of May at a location in Zimbabwe (latitude 19° South, longitude 30° East)

withdrawal at 7 and 8 am, then because of powerful solar radiation, it rises fairly fast from 9 am until 2 pm, despite some substantial hot water withdrawal. Between 2 pm and 5 pm, the rate of temperature rise slows down as the solar radiation tapers down, even though there is little hot water withdrawal during this period. After 5 pm, the combination of no radiation income, substantial water withdrawal and storage tank heat losses contribute to the tank temperature falling significantly. It is important to check that the maximum temperature in the storage tank does not exceed a critical design threshold (e.g. boiling point) over all weather conditions. For this system, the maximum storage tank temperature in October (hottest and highest radiation month) was determined to be below 80 °C, which is considered low enough to avoid boiling and associated system component failures.

1.6 Conclusion

Systematic methodologies are needed in order to make cost-effective decisions about choice and size of solar collector to employ in a solar water heating system. The *energy-per-dollar* metric as defined in this study, is one such instructive decision-making metric, as it includes all the important few collector attributes that influence life-cycle cost-effectiveness, i.e. collector *warranty life*; *energy output per unit area* and *cost per unit area*. In the study, a sample of SRCC-rated collectors, with differently-attractive attributes such as low cost-per-area, excellent efficiency curves or long warranty lives, were ranked using the energy-per-dollar metric. Flat plate type of collectors occupied the top four ranks, for the climatic conditions and load temperature under consideration. For the top-ranked collector (26.1 kWh/$), the collector area prescribed in the optimally-sized solar water heating system was

18 m^2/m^3 of hot water demand, the solar fraction 0.91, the storage-demand ratio 0.90 and the NPV was $5881/m^3$ of hot water demand. Although the method of this paper approach was applied only for a solar water heating application of specified operating temperature, at a specified location, it can be applied equally well for any other solar water heating application and at any other location.

References

1. Wbdg.org., Solar Water Heating | WBDG Whole Building Design Guide (2018). [online] Available at: https://www.wbdg.org/resources/solar-water-heating. Accessed 5 Jan 2018
2. Zetdc.co.zw., Tariffs | ZETDC (2018). [online] Available at: http://zetdc.co.zw/tariffs/. Accessed 5 Jan 2018
3. T. Hove, E. Manyumbu, G. Rukweza, Developing an improved global solar radiation map for Zimbabwe through correlating long-term ground- and satellite-based monthly clearness index values. Renew. Energy **63**, 687–697 (2014)
4. Zesa.co.zw., Why do we have Load Shedding? (2018). [online] Available at: http://www.zesa.co.zw/index.php/component/k2/item/17-why-do-we-have-load-shedding. Accessed 6 Jan 2018
5. Nama-database.org., National Solar Water Heating Programme—NAMA Database (2018). [online] Available at: http://www.nama-database.org/index.php/National_Solar_Water_Heating_Programme. Accessed 6 Jan 2018
6. K. Hudon, T. Merrigan, J. Burch, J. Maguire, Low-cost Solar Water Heating Research and Development Roadmap (2018) [online] Available at: http://www.ctgn.qc.ca/images/bulletins/bulletin_vol4no3/pdf/gb_solar_nrel_report.pdf. Accessed 6 Jan 2018
7. S. Sillman, The Trade-Off Between Collector Area, Storage Volume, And Building Conservation In Annual Storage Solar Heating Systems (1981). [online] Nrel.gov. Available at: https://www.nrel.gov/docs/legosti/old/907.pdf. Accessed 7 Jan 2018
8. A. Rushforth, Comparing Collectors Using SRCC Rating Data (2007). [online] Builditsolar.com. Available at: http://www.builditsolar.com/References/Ratings/CollectorCompare.htm. Accessed 7 Jan 2018
9. J.A. Duffie, W.A. Beckman, *Solar Engineering of Thermal Processes* (Wiley, Hoboken, NY, 2013), p. 399
10. Solar-rating.org., Solar Rating & Certification Corporation—Certification & Listing Directory (2018). [online] Available at: http://www.solar-rating.org/certification_listing_directory/. Accessed 7 Jan 2018
11. S.A. Klein, Calculation of flat-plate collector utilizability. Sol. Energy **21**(5), 393–402 (1978)
12. B. Gravely, Optimal Design in Solar Hot Water Systems—How Does Size Alter Performance? (2018) [online] http://www.solarhotwater-systems.com. Available at: http://www.solarhotwater-systems.com/optimal-design-in-solar-hot-water-systems-how-does-size-alter-performance/. Accessed 7 Jan 2018
13. T. Hove, E. Manyumbu, G. Rukweza, Developing an improved global solar radiation map for Zimbabwe through correlating long-term ground- and satellite-based monthly clearness index values. Renew. Energy **63**, 687–697 (2014)
14. M. Collares-Pereira, A. Rabl, The average distribution of solar radiation-correlations between diffuse and hemispherical and between daily and hourly insolation values. Sol. Energy **22**(2), 155–164 (1979)
15. T. Hove, J. Goettsche, Mapping global, diffuse and beam solar radiation over Zimbabwe. Renew. Energy **18**(4), 535–556 (1999)
16. B. Liu B, R. Jordan R, Daily insolation on surfaces tilted towards the equator. Trans. ASHRAE, 526–541 (1962)

17. C. Wit, *Simulation of assimilation, respiration and transpiration of crops* (Centre for Agricultural Publishing and Documentation, Wageningen, 1978), p. 148
18. T. Toy, A. Kuhaida, B. Munson, The prediction of mean monthly soil temperature from mean monthly air temperature. Soil Sci. **126**(3), 181–189 (1978)
19. R. Rankin, P.G. Rousseau, Sanitary hot water consumption patterns in commercial and industrial sectors in South Africa: impact on heating system design. Energy Convers. Manage. **47**(6), 687–701 (2006)
20. J.P. Meyer, A review of domestic hot water consumption in South Africa. R & D J. **16**(2000), 55–61 (2000)
21. T. Hove, N. Mhazo, Optimal sizing and economic analysis of low-cost domestic solar water heaters for Zimbabwe. J. Energy South Afr. **14**(4), 141–148 (2003)

Chapter 2
Assessment of Decentralized Hybrid Mini-grids in Sub-Saharan Africa: Market Analysis, Least-Cost Modelling, and Job Creation Analysis

A. Okunlola, O. Evbuomwan, H. Zaheer and J. Winklmaier

Abstract With a growing impetus to meet energy demand through decentralized hybrid mini-grids in rural and semi-urban locations in Sub-Saharan Africa (SSA), the need to accurately assess the market drivers, policy requirements and job creation impacts of this energy system typology within this region cannot be ignored. This work provides a techno-economic impact analysis of decentralized hybrid energy systems in selected locations in SSA. To optimally satisfy an electricity demand time-series for a year and minimize all cost components amortized over a period of 20 years, a least-cost modelling approach and tool is applied. An Employment Factor approach was used to calculate the direct employment impacts across the value chain of different hybrid mini-grid types. Additionally, the Leontief Inverse Input–Output model is used to determine the backward linkage economy-wide-jobs (gross jobs) created. The preliminary results show that the "Solar + Wind + Diesel + Battery" hybrid system (SWDB) has the lowest Levelized Cost of Electricity (LCOE), thus it provides the cheapest means of meeting the electricity demand in the modelled regions. However, the highest locally created direct and net employment impact in the model locations is provided by the "Wind + Battery" (WB) system. Two major sectors, manufacturing and agriculture have the largest number of gross jobs in the local economy for all decentralized hybrid systems analysed. This occurs due to higher linkages between these two sectors and the productive energy use in the area. Conversely, despite higher employment impacts obtained for WB, the cost and duration needed for

A. Okunlola (✉)
Institute for Advanced Sustainability Studies, Berliner Strasse 130, 14467 Potsdam, Germany
e-mail: ayodeji.okunlola@iass-potsdam.de

O. Evbuomwan
Ilf Beratende Ingenieure GmbH, Werner-Eckert-Strasse 7, 81829 Munich, Germany

H. Zaheer
Power and Technology Research LLC, Firkenweg 5, 85774 Unterföhring, Germany

J. Winklmaier
Technical University of Munich, Lichtenbergstrasse 4a, 85748 Garching, Germany

M. Mpholo et al. (eds.), *Africa-EU Renewable Energy Research and Innovation Symposium 2018 (RERIS 2018)*, Springer Proceedings in Energy,
https://doi.org/10.1007/978-3-319-93438-9_2

wind resource mapping and assessment serve as a major bottleneck to WB systems market access in the regions. The results of the sensitivity analysis suggest that by de-risking economic factors, such as discount rates, market access for decentralized renewable energy mini-grids can be improved in SSA.

Keywords Decentralized hybrid energy systems · Electricity access Job creation · Employment impact · Least-cost energy system modelling

2.1 Introduction

It has become acceptable that the motivation for a growth in renewable energy (RE) application globally is not only limited to reasons of energy security or reduction in CO_2 emissions from power generation sources, but also because of the positive employment impacts RE engenders [1]. This in turn supports the point that access to electricity through RE and economic development are strongly interlinked [2]. Countries located in Sub-Saharan Africa (SSA) have huge technologically feasible RE potentials greater than the average energy consumption needs of the sub-continent [3]. Nonetheless, the urban population in multiple countries in SSA remains underserved while many rural areas have little or no access to electricity. Although little literature exists to validate the correlations between electricity and employment generation in Sub-Saharan Africa (SSA), meta-analyses such as Daniel et al. show that investments in renewable-energy-based power generation provides more jobs per installed capacity (Jobs/MW) and per unit of energy generated (Jobs/MWh) over the operational lifetime of the power plant than the fossil-fuel-based power generation [1]. This applies for both centralized and decentralized electricity generation modes. Colombo et al. points out that the deployment of RE systems in decentralized modes of operation have the potential to create a larger amount of jobs per unit energy produced in comparison to conventional centralized methods of production and distribution of RE [4]. To justify that improved energy access enhances job creation, evidence from South Africa by Taryn Dinkelman established that household electrification driven by decentralized energy access increased employment in developing areas, particularly in micro-enterprise development [5].

2.2 Techno-economics and Job Creation

2.2.1 *Techno-economics of Mini-grids in Sub-Saharan Africa*

To assess the techno-economics of individual decentralized technologies, evaluating the technological appropriateness and the associated economic viability of each technology for a successful application are needed [6]. In their analysis in 2011, Szabó et al. showed the levelized cost of electricity (LCOE) calculated for local mini-grid photovoltaic (PV) systems in Africa ranges from approximately 0.2 to 0.55 \$/kWh [7, 8]. Similarly, Deichmann et al. in 2010 showed that the average LCOE for wind-based mini-grids in Africa ranges between 0.144 and 0.288 \$/kWh [9]. By comparing[1] LCOE results from [8] to [9], wind energy can be considered to be more favourable for mini-grid application in Africa due to its lower LCOE. However, low wind speeds observed across SSA serve as a major limiting factor to its wide application for power generation on the sub-continent [10].

2.2.2 Renewable Energy Employment Effects

The value chain of a RE system consists of economic activities which trigger direct, indirect, and induced employment effects [11]. Direct employment effects refer to un-intermittently created jobs in the value chain such as manufacturing, construction and installation (C&I) as well as operations and maintenance (O&M) jobs. Indirect employment effects accrue from industries which are linked to the renewable energy value-chain. Besides direct and indirect employment effects, there are also induced employment effects which occur as a result of income distribution effects [12]. For analysing the total employment effects of RE systems, the terms gross and net effects are used. Gross employment effects refer to positive job creation in an economy as a result of investment in the RE sector. Net employment effect, on the other hand, shows both the positive job effects (direct, indirect and induced) and negative job effects (job losses in the conventional energy sector). Ürge-Vorsatz et al. points out that in developing regions, such as SSA, the positive employment effects are often sufficient to be analysed because these effects often start at a low level that they almost always lead to positive gains [13].

There are many models and tools used in calculating job creation effects of RE varying on the basis of complexity, data requirements, and computational assumptions. However, two main approaches exist in determining gross employment effects. The approaches are; (1) the employment factor (EF) method, and (2) the gross input-output (IO) approach. EFs indicate the number of jobs generated per unit of installed capacity (MW) or energy produced (MWh) [14], while the gross IO model is applied to assess both the gross and net employment effects of RE systems [12] (Fig. 2.1).

[1]The techno-economic comparison does not take into consideration the huge drop in the system cost of solar PV from the years 2010 to 2017 (till date).

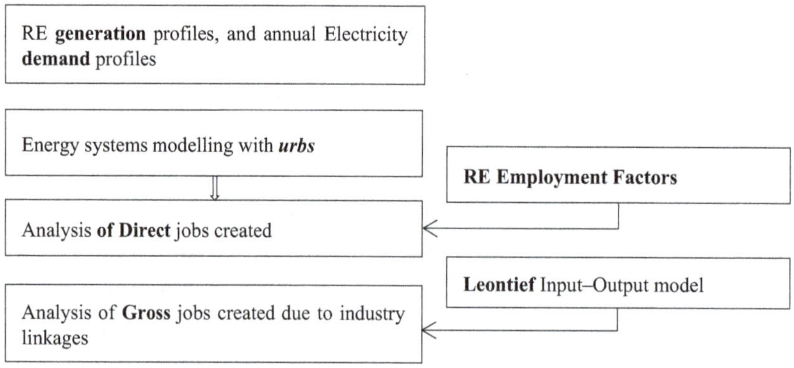

Fig. 2.1 Assessment methodology adopted in the study

2.3 Methodological Approach

In this study, the least cost annual energy supply of a semi-urban location in SSA is computed using urbs, a linear optimization model for distributed energy systems developed at the Institute of Renewable and Sustainable Energy Systems, Technical University Munich, Germany. Details of the urbs model and the opitimization logic have been reported in the documentation [15]. The model is used for the techno-economically optimized sizing of ten decentralized hybrid mini-grid scenarios and systems combinations and thus comparing them. Scenarios considered are shown in Table 2.1.

In some scenarios battery storage is added to balance RE supply fluctuations, while the already installed diesel generator served as the conventional backup system. Techno-economic simulation parameters and the energy flow strategy used in the model are adopted from Bertheau et al. [16]. Input technical and economic data for the mini-grid components are provided by ILF BERATENDE INGENIEURE GmbH. An appliance utilization and daily power consumption analysis for 10 energy demand clusters was conducted in the region. The results obtained during this process are used in a Monte-Carlo time-of-use simulation to determine the annual hourly

Table 2.1 Hybrid mini-grid scenarios and configurations used in the analysis

Scenario	Abbreviation
Diesel only	Base
Solar diesel	SD
Solar + diesel + battery	SDB
Solar + wind + diesel	SWD
Solar + wind + diesel + battery	SWDB
Solar + wind + battery	SWB
Solar + battery	SB
Wind + battery	WB
Wind + diesel	WD
Wind + diesel + battery	WDB

electricity demand profile. Annual hourly solar radiation used is derived from NREL and wind speed data are obtained from the Global Wind Atlas [17, 18]. Adiabatic terrain index in Counihan was used to adjust the wind speeds obtained according to the terrain surface roughness of the area [19]. The calculations for adjusted wind speeds of a typical semi-urban terrain type are shown in Mayr [20]. The methodology applied in Bertheau et al. is used to calculate the LCOE for all scenarios at 15% weighted average cost of capital (WACC). The LCOE gives the cost-optimized system configurations for a project period of 20 years. The system sizing results from urbs are used to compute the employment impacts for all scenarios. Replacement and degradation costs are considered to be of negligible effect. Also, system stability simulations are not conducted in this exercise. The direct jobs created are calculated applying the EF methodology in Rutovitz et al. [21]. The country's productivity level obtained from KILM [22] is used to adjust the EF for the region in estimating the direct jobs created locally. The backward-linkage jobs (gross jobs) created in different industries in the region are calculated using the Leontief IO model [23]. The regional IO table required for the IO model is built applying [24–26].

2.4 Results and Discussion

SDB, SDW, SDWB, have the lowest LCOE, thus they provided the cheapest cost required to satisfy the electricity demand in the modelled region (Fig. 2.2). The three scenarios have marginally equal LCOEs because the percentage of energy supply from diesel generators in all three scenarios is approximately 72% (LCOEs are rounded up to 3 decimal places for simplicity; slight LCOE differences are observed beyond this range). The influence of battery storage when a hybrid mini-grid is optimized with a diesel generator is trivial. SB's LCOE is 1.61 $/kWh cheaper than WB because of solar PV's significantly lower investment cost and

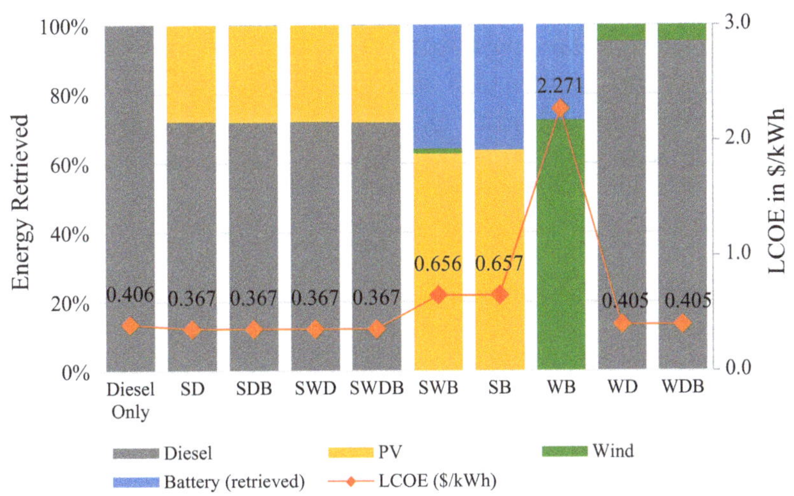

Fig. 2.2 LCOE and energy retrieval by supply source for all scenarios

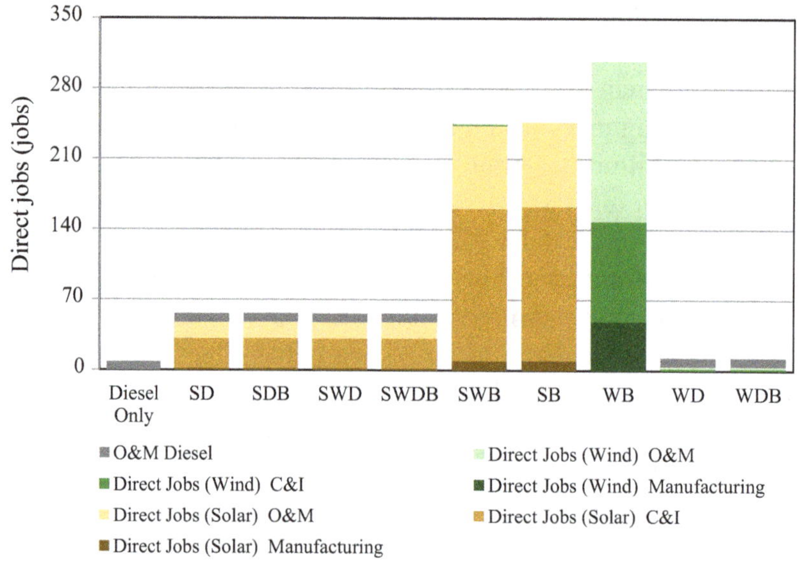

Fig. 2.3 Direct jobs created in Manufacturing, C&I, and O&M for all scenarios

higher on-site capacity factor (CF). WB had the highest LCOE (2.27 \$/kWh), thus, the most expensive mini-grid scenario. The wind turbine has a significantly lower CF than solar due to the poor wind speed (<4 m/s) in the area. This is typical in most locations in SSA [9]. Therefore, scenarios where the hybrid mini-grid has a system configuration of solar with wind, power generation from wind has a near-zero effect on the annual energy supply.

On the other hand, in spite of WB scenario providing the highest LCOE over the lifetime of the project, WB generates peak direct employment (Fig. 2.3) and gross employment effects (Fig. 2.4). This high employment effect is due to a high rated capacity of the installed wind turbine as compared to solar in SB which influences C&I and O&M jobs. This therefore results into huge investment flow from the mini-grid installation into the local economy. The created direct manufacturing jobs are negligible in all ten scenarios because of a low percentage of in-country RE equipment manufacturing. The highest gross employment effects are obtained in the trading, agriculture, and maintenance and repair industries for all mini-grid combinations (Fig. 2.4). These industries create high employment linkage to electricity supply in the local economy as identified from the IO model.

2.5 Sensitivity Analysis

A sensitivity analysis was conducted to investigate the impact of wind turbines optimized for low wind speeds, capital expenditure (CAPEX) reduction, and WACC on the LCOE and direct jobs created in the SWDB mini-grid scenario. The scenario is chosen because of its previous low LCOE result. A new location with an annual demand profile similar to a typical rural area in SSA and also allowing for a

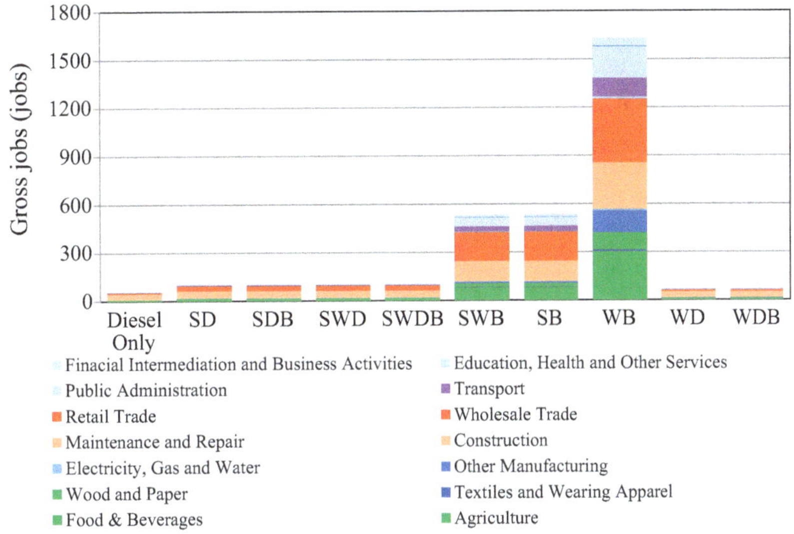

Fig. 2.4 Gross jobs created industries in the local economy for all scenarios

locally manufactured (DIY) low-wind-speed Piggott turbine in [19] to be built is chosen. The results obtained represented in Figs. 2.5 and 2.6 show the effect of WACC changes on LCOE, the corresponding direct job creation, and also the effect of RE CAPEX reduction. Three distinct results are obtained from the sensitivity analysis:

1. Low WACC significantly increases direct jobs and reduces LCOE (Fig. 2.5). This occurs because lower financing cost ensures a reduction in LCOE where an annual levelized capital cost dominates the average annual operating cost, hence increasing the installation of RE technologies in the SWDB mini-grid. The final result is an even higher increase in the number of direct jobs created due to higher investment in RE.
2. At lower WACC, more direct jobs are created for wind turbines optimized for low wind speeds than those created for Solar PV (Fig. 2.5). Optimizing wind turbines for low wind speed increases the CF of the wind turbines in rural areas. Therefore, with low financing costs, more installations of low- wind-speed turbines can be achieved, hence a positive impact of direct local jobs.
3. Solar LCOE is more sensitive to CAPEX reduction, but higher job creation sensitivities can be seen for small-wind turbines as the CAPEX reduces (Fig. 2.6). Solar generates a lower LCOE as compared to DIY wind turbines in rural SSA because it requires lower investment costs as CAPEX reduces. However, from 75 to 90% CAPEX reduction, wind DIY begins to generate more direct jobs than solar, despite having a higher LCOE. This occurs because of lower WACC inducing reduction in investment costs for the DIY turbines coupled with an improved CF of the wind turbines which are optimized to perform better at low wind speeds. This combined effect results in more turbines being installed and thus has a higher positive effect on direct job creation in the value chain.

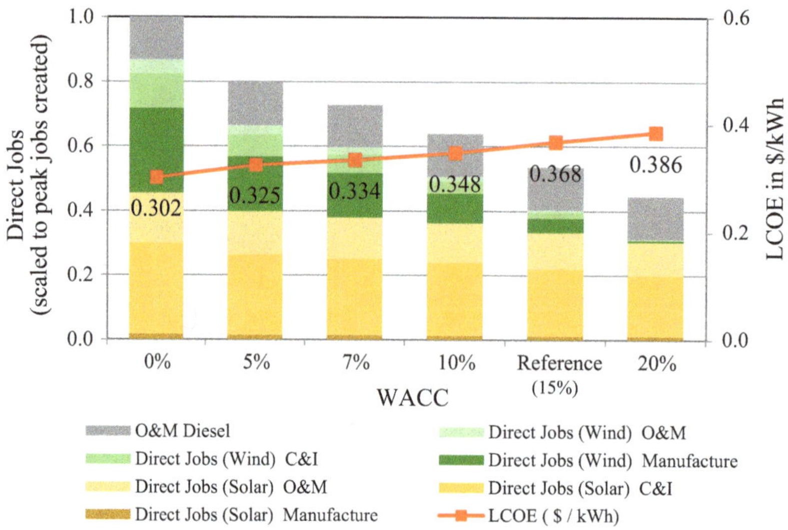

Fig. 2.5 Effect of changes in WACC on direct jobs and LCOE for SWDB

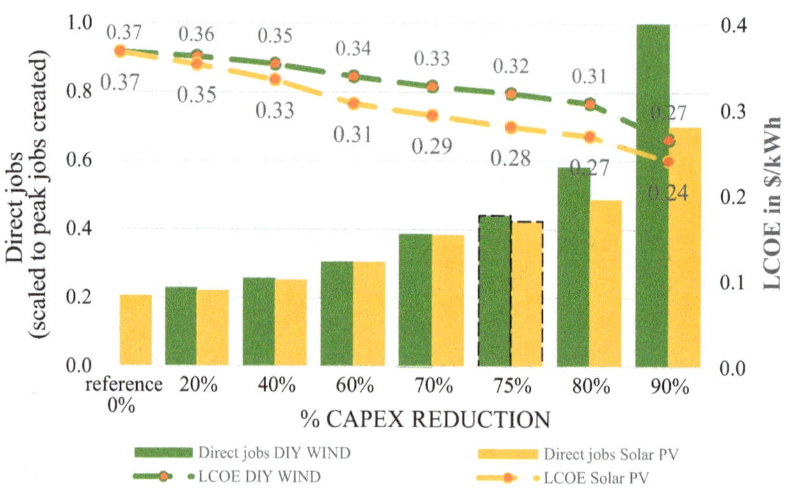

Fig. 2.6 Effect of CAPEX Reduction on LCOE and Direct Job for SWDB

2.6 Conclusions and Recommendations

Results obtained in this study show an optimized configuration for a decentralized hybrid mini-grid in SSA based on the cost of generating electricity and the jobs it creates. Small wind turbines optimized for lower wind speeds are better suited for hybrid combination with solar PV than commercially available wind turbine technologies in locations where annual average wind speeds are below 4 m/s. Notwithstanding the high LCOE of WB mini-grid, the high employment impacts it creates makes it a probable option for rural electricity supply in SSA. Bottlenecks like high cost and duration of site-specific wind resource assessment present a major challenge to the growth of decentralized wind systems in SSA despite the significant potentials for hybridization. In terms of scalability, the higher modularity of solar PV ensures it remains a major driver of hybrid mini-grids in SSA. By de-risking economic factors such as discount rate, the WACC can be reduced; hence improvement in market access for decentralized RE hybrid mini-grids can be realized in SSA. To achieve desired positive job creation impacts, effective policy measures such as those being developed within the COBENEFITS[2] project [27] should be considered and applied; these policy measures place specific emphasis on the multiple benefits of increased share of RE in the energy mix [28]. Further studies should focus on developing models to broadly investigate the induced and net employment effects of decentralized renewable energy systems in Africa with respect to the techno-economics, taking into account the distribution costs and system stability requirements.

Annexes

See Fig. 2.7, Tables 2.2, 2.3 and 2.4

Gross IO model principal equations:

$$X = [1 - A]^{-1} * Y \tag{2.1}$$

$$Totaljobs_x = \sum \left(\left[\frac{\text{Wi}}{\text{\$output}} \right] \left[\left((1 - \text{A})^{-1} \right) \right] [\text{Totalcosts}_\text{s}] \right)_{\text{Ti}} \tag{2.2}$$

where:

X = *output multiplier matrix;*
A = *Technical* (*input*) coefficient matrix
$[1 - A]^{-1}$ = Leontief inverse matrix

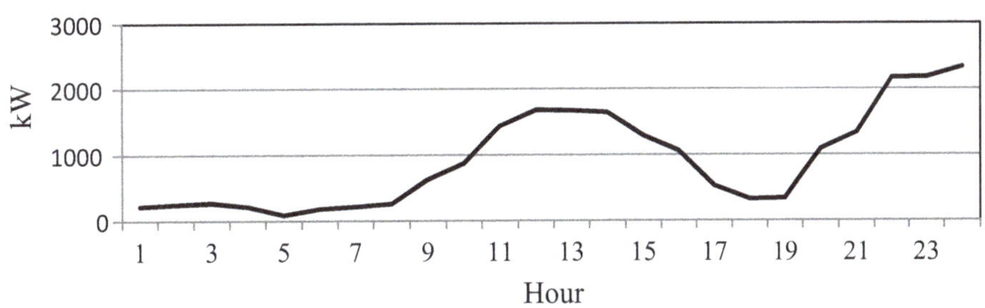

Fig. 2.7 An average hourly energy demand profile for a day obtained from the Monte-Carlo simulation

[2]The COBENEFITS project at the Institute for Sustainability Studies Potsdam, Germany is part of the International Climate Initiative (IKI) supported by the German Federal Ministry for the Environment, Nature Conservation, Building and Nuclear Safety (BMUB).

Table 2.2 System sizing results

Scenario	Rated power solar PV (kW)	Rated power wind (kW)	Battery (kWh)	LCOE ($/kWh)
Base	0	–	–	0.406
SD	2202	–	–	0.367
SDB	2207	–	12	0.367
SWD	2202	–	–	0.367
SWDB	2207	–	12	0.367
SWB	11,241	215	34,351	0.656
SB	11,384	–	34,936	0.657
WB	–	32,550	142,846	2.271
WD	–	452	–	0.405
WDB	–	452	–	0.405

Table 2.3 Employment impact results

Scenario	Direct jobs			Gross jobs
	Manufacture	C&I	O&M	
Base	–	–	–	190
SD	2	30	16	220
SDB	2	30	16	220
SWD	2	30	16	220
SWDB	2	30	16	220
SWB	10	152	84	780
SB	10	153	84	780
WB	49	100	159	1950
WD	1	1	2	200
WDB	1	1	2	200

Table 2.4 Employment factors for the location

Technology	(Jobs/MW)			FUEL (jobs/GWh)
	Manufacture	C&I	O&M	
PV	8.44	13.46	7.34	0.00
Wind	7.47	3.06	4.90	0.00
Diesel	1.22	2.08	1.96	2.94

Y = Total cost of the minigrid system

$Totalcosts_s$ = *(investment cost + variable costs) in a scenario*

Ti = *benchmark year. In this case the benchmark year is* 2017;

Wi = *workers per industry*

Total $jobs_x$ = *Total gross jobs in a minigrid scenario*

See Tables 2.5 and 2.6

Table 2.5 Economic parameters used in the techno-economic analysis

Input parameter	Parameter value
Weighted cost of capital	15%
Diesel generator	
Initial investment cost	240 $/kW
Fixed cost	$30 $/kW/year
Variable cost	0.03 $/kWh
Fuel cost	$0.11 $/kWh
Solar photovoltaic	
Initial investment cost	1600 $/kW
Fixed cost	2% of Investment/year
Variable cost	0.03 $/kWh
Economic lifetime	20 years
Wind turbine	
Initial investment cost (commercial)	2000 $/kW
Initial investment cost (DIY)	3005 $/kW
Fixed cost	2% of Investment/year
Variable cost (commercial)	0.02 $/kWh
Variable cost (DIY)	0.05 $/kWh
Local content requirement (commercial)	30%
Local content requirement (DIY)	90%
Economic lifetime	20 years
Battery storage	
Initial investment cost (power)	500 $/kW
Initial investment cost (energy)	300 $/kWh
Fixed cost (power)	2% of Investment cost/year
Fixed cost (energy)	10 $/kWh/year
Variable cost (power)	0.02 $/kW
Variable cost (energy)	0.02 $/kWh
Economic lifetime	10 years

Table 2.6 Technical Parameters the Techno-Economic Analysis

Input parameter	Parameter value
Diesel generator	
Full load efficiency	30%
Minimum load efficiency	25%
Minimum load	25% of rated capacity
Solar PV system	
Module type	Polycrystalline
Module efficiency	16.7%
Rated module power	325 W at Standard test conditions
Solar inverter efficiency	98%
Life span	20 years
Wind turbine	
Wind turbine manufacturer	Bergey
Cut–in speed (commercial)	3 m/s
Rated speed (commercial)	11 m/s
Cut–in speed (DIY)	2 m/s
Rated speed (DIY)	5 m/s
Hub height	50 m
Life span	20 years
Power coefficient (Cp)	0.33 (Commercial) 0.29 (DIY)
Battery storage	
Type	Lithium-ion
Input efficiency	95%
Output efficiency	93%
Initial state of charge (SOC)	50%
Life span	10 years

References

1. D.M. Kammen, K. Kamal, M. Fripp, *Putting Renewables to Work: How Many Jobs Can the Clean Energy Industry Generate*? RAEL Report, University of California Berkeley (2017), http://rael.berkeley.edu/old_drupal/sites/default/files/very-old-site/renewables.jobs.2006.pdf. Accessed 17 Jan 2017
2. IRENA, *Solar PV in Africa: Costs and Markets* (2016), http://www.irena.org/publications/2016/Sep/Solar-PV-in-Africa-Costs-and-Markets. Accessed 10 Jan 2017
3. S. Hermann, A. Miketa, N. Fichaux, *Estimating the Renewable Energy Potential in Africa,* IRENA-KTH working paper (International Renewable Energy Agency, Abu Dhabi, 2014), http://www.irena.org/-/media/Files/IRENA/Agency/Publication/2014/IRENA_Africa_Resource_Potential_Aug2014.ashx. Accessed 10 May 2017
4. E. Colombo, S. Bologna, D. Masera, *Renewable Energy for Unleashing Sustainable Development*, p. 234. ISBN: 9783319002842 (2013)

5. T. Dinkelman, *The Effects of Rural Electrification on Employment: New Evidence from South Africa* (2010), https://www.princeton.edu/rpds/papers/dinkelman_electricity_0810.pdf. Accessed 17 Jan 2017
6. M. Jamil, S. Kirmani, M. Rizwan, Techno-economic feasibility analysis of solar photovoltaic power generation: A review. Smart Grid Renew. Energy **3**, 266–274 (2012). https://doi.org/10.4236/sgre.2012.34037
7. A Manual for the Economic Evaluation of Energy Efficiency and Renewable Energy Technologies. National Renewable Energy Laboratory (NREL). http://www.nrel.gov/docs/legosti/old/5173
8. S. Szabó, K. Bódis, T. Huld, M. Moner-Girona, Energy solutions in rural Africa: Mapping electrification costs of distributed solar and diesel generation versus grid extension (2011)
9. U. Deichmann, C. Meisner, S. Murray, D. Wheeler, *The Economics of Renewable Energy Expansion in Rural Sub-Saharan Africa*. Policy Research Working Paper; No. 5193 (World Bank, Washington DC. © World Bank, 2010), https://openknowledge.worldbank.org/handle/10986/19902. Accessed 13 Feb 2017
10. A.D. Mukasa, E. Mutambatsere, Y. Arvanitis, T. Triki, *Development of Wind Energy in Africa*, Working Paper Series No. 170 (African Development Bank, Tunis, Tunisia, 2013)
11. B. Breitschopf, C. Nathani, G. Resch, *Methodological guidelines for estimating the employment impacts of using renewable energies for electricity generation* (2012), http://iea-retd.org/wp-content/uploads/2012/12/EMPLOY-Guidelines.pdf. Accessed 13 Oct 2017
12. S. Borbonus, *Generating Socio-economic Values from Renewable Energies: An Overview of Questions and Assessment Methods*. Institute for Advanced Sustainability Studies Potsdam, Working paper (2017), https://doi.org/10.2312/iass.2017.016
13. D. Ürge-Vorsatz, S.T. Herrero, N.K. Dubash, F. Lecocq, Measuring the Co-Benefits of climate change mitigation. Annu. Rev. Environ. Resour (2014). https://doi.org/10.1146/annurev-environ-031312-125456
14. IRENA, CEM, *The Socio-economic Benefits of Large-Scale Solar and Wind*, an econValue report (2014), http://www.irena.org/documentdownloads/publications/socioeconomic_benefits_solar_wind.pdf. Accessed 03 Mar 2017
15. urbs. A linear optimisation model for distributed energy systems. http://urbs.readthedocs.io/en/latest/
16. P. Bertheau, C. Cader, H. Huyskens, P. Blechinger, The influence of diesel fuel subsidies and taxes on the potential for solar-powered hybrid systems in Africa. *Resources* 2015 **4**, 673–691 (2015)
17. NREL, PVWatts Calculator. http://pvwatts.nrel.gov/. Accessed 24 Jan 2017
18. Global Wind Atlas 2.0, A free, web-based application developed, owned and operated by the Technical University of Denmark (DTU) in partnership with the World Bank Group, utilizing data provided by Vortex, with funding provided by the Energy Sector Management Assistance Program (ESMAP). https://globalwindatlas.info/. Accessed 24 Mar 2017
19. J. Counihan, Adiabatic atmospheric boundary layers: A review and analysis of data from the period 1880–1972. Atmos. Environ. **9**, 871–905 (1975). https://doi.org/10.1016/0004-6981(75)90088-8
20. M. Mayr, Feasibility analysis of power supply by small scale wind turbines in urban, semi-urban and rural districts of Zimbabwe. Master's thesis, Technical University Munich, 2017
21. J. Rutovitz, S. Harris, *Calculating Global Energy Sector Jobs: 2012 Methodology*. Prepared for Greenpeace International by the Institute for Sustainable Futures (University of Technology, Sydney Australia, 2012), https://opus.lib.uts.edu.au/bitstream/10453/43718/1/Rutovitzetal2015Calculatingglobalenergysectorjobsmethodology.pdf. Accessed 17 Jan 2017
22. International Labour Organization. Key Indicators of the Labour Market (KILM) statistical database (2017). http://www.ilo.org/global/statistics-and-databases/research-and-databases/kilm/lang–en/index.htm
23. W. Leontief, Quantitative input and output relations in the economic systems of the United States. Rev. Econ. Stat. **18**(3), 105–125 (1936). https://doi.org/10.2307/1927837

24. M. Lenzen, K. Kanemoto, D. Moran, A. Geschke, Mapping the structure of the world economy. Environ. Sci. Tech. **46**(15), 8374–8381 (2012). https://doi.org/10.1021/es300171x
25. M. Lenzen, D. Moran, K. Kanemoto, A. Geschke, Building Eora; A Global Multi-regional Input-Output Database at High Country and Sector Resolution, Economic Systems Research, 25:1, 20 49. (2013), https://doi.org/10.1080/09535314.2013.769938
26. A.P. Thirlwall, in *Growth and Development with Special Reference to Developing Countries*. 8th edn. (Palgrave Macmillan, New York, 2006) (The Review of Economics and Statistics)
27. COBENEFITS: Mobilising the Multiple Opportunities of Renewable Energies, 2016–2020. https://www.iass-potsdam.de/en/research/renewables
28. S. Helgenberger, M. Jänicke, in *Mobilizing the Co-benefits of Climate Change Mitigation Connecting Opportunities with Interests in the New Energy World of Renewables.* Institute for Advanced Sustainability Studies Potsdam, Working paper (2017)

Chapter 3
Feasibility Study of Linear Fresnel Solar Thermal Power Plant in Algeria

Hani Beltagy, Sofiane Mihoub, Djaffar Semmar and Noureddine Said

Abstract Clean renewable electric power technologies are important in human life, a great number of thermal solar power plants with different configurations are being considered for deployment all over the world. In this work, we propose a feasibility study of concentrated solar power plant to be set up in different sites of Algeria. It is essential that the plant design will be optimized to each specific location. Among the CSP technologies, we will emphasize on the Fresnel solar power plants at different areas of Algerian Sahara. These areas have been chosen for comparison by shifting the plant to different locations; namely Hassi R'mel, Tamanrasset, Beni-Abbes, and El Oued. Direct Normal Irradiance (DNI), solar field surface, block number, the block surface, block panels' number, absorber surface, and finally thermal power losses in the absorber are the key parameters for optimization and performance evaluation. The calculation results have been depicted for each site. Indeed, the calculation of performance varies from one site to another with DNI mean values of 788.4, 698.7, 671.7, and 636 W/m^2, respectively for Tamanrasset, Beni-Abbes, El Oued and Hassi R'mel sites. The surface of solar field, block number, absorber surface and power loss have been also evaluated for the same sites.

H. Beltagy (✉)
Mechanical Engineering Department, Faculty of Technology,
Blida University, Blida, Algeria
e-mail: hani.beltagy@gmail.com

S. Mihoub
Annexe de Sougueur, Faculté de Sciences de La Matière, Tiaret University, Tiaret, Algeria
e-mail: mihoubsofiane@yahoo.fr

D. Semmar
Renewable Energies Department, Faculty of Technology, Blida University, Blida, Algeria
e-mail: djaffarsemmar@yahoo.fr

N. Said
Renewable Energy Development Center (EPST/CDER), Bouzareah, Algiers, Algeria
e-mail: saidnoureddine@hotmail.com

M. Mpholo et al. (eds.), *Africa-EU Renewable Energy Research and Innovation Symposium 2018 (RERIS 2018)*, Springer Proceedings in Energy,
https://doi.org/10.1007/978-3-319-93438-9_3

Keywords Solar energy · Solar concentrator mirrors · Fresnel solar thermal power plants · Performance · Thermoelectric plants

3.1 Introduction

It is universally acknowledged that two of the key technological and economic challenges of the 21st century are energy and environment [1]. Consequently, considerable efforts are being made to effect a gradual transition from systems based on fossil fuels to those based on renewable energy (RE). Electricity generation from solar energy is currently one of the main research areas in the field of renewable energy. In the case of Algeria; the newly adopted version of the National Renewable Energy Program offers the country the possibility to integrate 27% of renewable energy in the national energy mix. Preservation of fossil re-sources; diversification of electricity production and contribution to sustainable development are among challenges that face the country nowadays [2]. It is recognized that solar-thermal energy can play a useful role in generating electrical power despite concerns regarding cost as the thermal source is accessible and ubiquitous [3]. The use of linear Fresnel solar power plant for electricity production, or heat supply is one of the most attractive solutions for developing countries with high sunshine because of the accessible level of technology involved [4].

In this study, our main purpose is to describe Fresnel concentrator solar plant characteristics, through defining its different performances, so that influence of changing location and climatic conditions can be seen on plant cost-effectiveness and productivity for each location i.e. Hassi R'mel, Beni-Abbes, El Oued, and Tamanrasset for Central, West, East and South of Algeria respectively.

3.2 Plant Description

The solar power plant chosen for this study is a 5 MW electric Fresnel concentrator plant which is technically similar to the German Novatec solar plant set up in Calasparra site in Spain (same make and model) (Fig. 3.1).

The design parameters of the power plant are shown in Table 3.1 (Novatec solar) [5].

3.3 Specification

In this section, we have drawn up a specification in which plant characteristics, namely Direct Normal Irradiance (DNI), solar field surface, block number, have been defined (it should be noted that a block contains reflecting mirrors, absorber

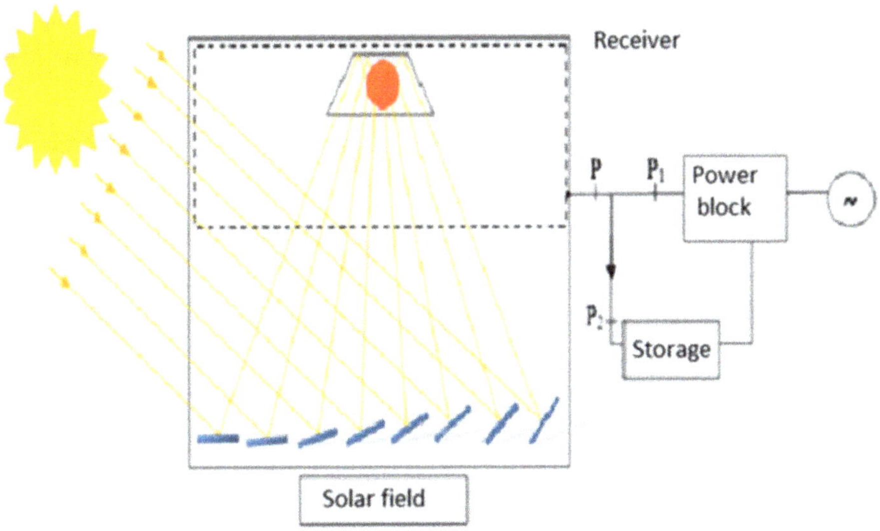

Fig. 3.1 Power plant layout

Table 3.1 Power plant characteristics

Mirror field	Solar field size Solar field length Net opening surface Field width Reflector surface Panel length Reflector length	21,571 m^2 806 m 18,489 m^2 16.56 m^2 513.6 m^2 44.8 m^2 0.75 m
Receiver	Tube diameter Receiver diameter Length reflector-absorber	0.07 m 0.6 m 7 m
Power block	Block efficiency Inlet temperature Outlet temperature Power at the generator output	35% 140 °C 270 °C 5 MW electric

tube, and steam separator). We have also defined the block surface, block panels' number, absorber surface, and finally thermal power losses in the absorber.

A. Sizing calculation

The generator output power is equal to: P_g = **5 MW$_{\text{electric}}$**.

In order to supply a small city of 5000 inhabitants, whose daily consumption is **5 kWh** per family of electricity, each family is composed of **5 persons**.

B. Distribution of powers per activity

Polyclinic = **0.2 MW**, factory = **1 MW**, Administration = **0.5 MW**, other = **0.3** MW. The sum is **2 MW**, so it remains **3 MW** for inhabitants: this is the peak power needed by these places.

Therefore, inhabitants number = **1000 houses**, or: x = 1000 * 5 = **5000 people**
So, the consumption of the city is **5 MWe,** it is distributed as follows:

- Polyclinic = 0.2 MW
- Factory = 1 MW
- Administration = 0.5 MW
- Inhabitants = 3 MW
- Others = 0.3 MW

The solar multiple is taken as equal to **1.6** which corresponds to **1.5 h** of storage. The selected locations are:

- West → Beni-Abbes
- Center → Hassi R'Mel
- East → El Oued
- South → Tamanrasset

Table 3.2 represents site meteorological data.

C. Calculation of plant characteristics

- DNI_{avg} calculation (direct normal irradiance)

$$DNI_{avg} = \frac{DNI_{annual}}{\text{Insolation period}} = \frac{2236}{3200} = 0.698\ \text{kW/m}^2 \times 10^3$$

= **698.7** W/m^2 for Beni-Abbes and the same calculation is done for the other locations.

- Calculation of power

$$\left.\begin{array}{r}\text{Generator efficiency} = 9.5\% \\ \text{Turbine efficience} = 38\% \\ \text{Parasitics efficiency} = 98\%\end{array}\right\}\text{we will obtain}$$

$$\eta_{bloc} = 0.95 * 0.38 * 0.98 = 0.35 = 35\%$$

Table 3.2 Site meteorological data [6]

Site selected	Beni-Abbess	Hassi R'mel	El Oued	Tamanrasset
DNI_{annual} (kWh/m^2 year)	2236	2035.5	2149.5	2759.4
$T_{amb,\ avg}$ (°C)	22	22.4	21.4	22.7
Insolation duration (h/year)	3200	3200	3200	3600
DNI_{avg} (W/m^2)	698.7	636	671.7	788.4

$$p_1 = \frac{p_g}{\eta_{bloc}} = \frac{5}{0.35} = 14.28\ \mathrm{MW_{thermal}}$$

$P = P_1 + P_2 = SM * P_1$, with SM is the solar multiple = 1.6
$P = 1.6 * 14.28 =$ **22.84 MW$_{\mathbf{thermal}}$**
So: $P_2 = 22.84 - 14.28 =$ **8.56 MW$_{\mathbf{thermal}}$**
All efficiencies and powers are shown in Table 3.3.

- Calculation of field surface for each location

If receiver efficiency = **85%** we will obtain: Optical efficiency = **45%**
So, global efficiency is obtained:

$$\eta_{global} = 0.85 * 0.45 = 0.38 = \mathbf{38\%}$$

$\eta_{global} = \frac{P}{A*\mathrm{DNI}} \Rightarrow A = \frac{P}{\eta_g*\mathrm{DNI}} = \frac{22.84\times10^3}{0.38*0.698} = 86,000\ \mathrm{m}^2$
For Beni-Abbes, and the same calculation is done for the other locations.
"A" represents field surface.

- Calculation of block number and surface of each block

The size of (Novatec) plant solar field is S = **21,571 m$^{\mathbf{2}}$**
This plant contains **2** blocks, each block has a power of **0.7 MW** electric

$\frac{A}{S} = \frac{86,000}{21,571} = 3.98$ is the number of block for a power of **0.7 MW** electric

$3.98 * 2 = 8$ blocks is the number of block for a power of **1.4 MW** electric
"A" is the block surface.
The surface of a block is given by:

$A = \frac{86,000}{8} = 10,750\ \mathrm{m}^2$ for Beni-Abbes and the same calculation is done for the other locations.

Table 3.3 Efficiencies and power [5]

Generator output	5 MW (%)	Global efficiency	38%
Generator efficiency	95	Solar multiple	1.6
Turbine efficiency	38	Number of storage hours	1.5 h
Accessory efficiency	98	Input temperature	140 °C
Power block efficiency	35	Output temperature	270 °C
Optical efficiency	45	Power P_1	14.28 MW$_{th}$
Receiver efficiency	85	Power P_2	8.56 MW$_{th}$

Table 3.4 Plant characteristics in each site

Selected locations	BA	Hassi R'mel	El Oued	Tam
DNI_{avg} (W/m^2)	698.7	636	671.7	788.4
Solar field surface	86,000	94,000	89,000	76,000
Block number	8	8	8	7
Block surface (m^2)	10,750	10,804	10,853	10,857
Unit number of a block	21	21	21	21
Actual absorber surface (m^2)	564.48	564.48	564.48	564.22
Total surface of absorbers (m^2)	4515.8	4911	4628.7	3949.5
Lost power (W)	2.93×10^6	3.19×10^6	3×10^6	3.5×10^6

- Calculation of actual absorber surface

Absorber unit surface is: $L * D$
L represents unit length = **44.8 m**, D represents receiver diameter = **0.6 m**
$A_{abs;\ unit} = 44.8 * 0.6 = \mathbf{26.88\ m^2}$
Each block contains a unit whose surface is 513.6 m^2

$$\text{Unit Number} = \frac{\text{Block surface}}{\text{Unit surface}}$$

$$\frac{10,750}{513.6} = 21\,\text{units}$$

So: $A_{abs\ actual}$ = Number of units multiplied by unit absorber surface.
= 21 * 26.88 = **564.48 m^2** is the absorber surface in a 0.7 MW block.

Total surface of absorbers is calculated by multiplying the result of block number by the actual absorber surface.

$S_{Total} = 8 * 564.48 = \mathbf{4515.8\ m^2}$ for Beni-Abbes and the same calculation is done for the other locations.

- Calculation of lost power

$P_{loss} = U_L * A_{abs\ actual} * \Delta T$
We have: T_{in} = **140 °C**, T_{output} = **270 °C**, U_L = **5 W/m^2 °C**

$P_{loss} = 5 * 4515.8 * 130 = \mathbf{2.93 * 106\ W}$ for Beni-Abbes and the same calculation is done for the other locations. Final results are shown in Table 3.4

Table 3.5 Comparative study

Selected sites	Hassi R'mel	Tamanrasset	Beni-Abbess	El Oued
DNI_{avg} (W/m^2) [M1]	636	788.4	698.7	671.7
Power (MW) [MI]	5	5	5	5
Power (MW) [SAM]	4.74	5.01	5.14	45.43
Error (%)	5.12	0.37	2.7	7.9

3.4 Results Analysis

As a comparison, we want to check the results found in the first part, then validate them through drawing up specifications in which different plant performances are defined, while using a global calculation method. These performances have been calculated, through using a DNI calculated on the basis of annual normal irradiation ratio to sunshine hours. We propose to verify hereafter these results with those found by SAM simulation software [7]. Simulation of energy performance is carried out in another study using a SAM simulation software, as well as the techno-economic study [8]. A comparative study is made while taking a representative day of the year (21st June, the longest/hottest), we took the DNI values, and then recalculate the power output to adjust the results for each site.

The results are shown in Table 3.5.

M1 shown in the table depicts the first calculation method (the global method set forth in the specifications). According, this comparative study aiming at the verification of the results obtained in the global method with those calculated by SAM simulation software, we notice that the output power varies from 4.74 to 5.43 MW, so we can see that there is a slight difference between powers calculated by both methods, and according to the calculations show a difference of less than 8%. The results found by the global method used for the cases studied here are quite satisfactory.

3.5 Conclusion

The study we carried out on Linear Fresnel Concentrator Solar Power Plant allowed us to understand the operating system of this type of plant, as well as the parameters affecting its operation and performances. We have also noticed that the blocking and shadowing effects between mirrors and the cosine effect represent a great obstacle for the power plant cost-effectiveness and productivity. According to this study, we were able to notice the importance of a good choice of the power plant site, as each site is characterized by its direct radiation, ambient temperature, sunshine hours, wind speed, latitude, height above sea level, and other factors which play a key role in the power plant cost-effectiveness and productivity.

This is clearly shown in the results obtained according to which the plant different parameters and characteristics vary when it is shifted in different sites.

According to the results obtained for the plant different characteristics in the 4 sites, we can see that Algeria has great opportunities to choose this kind technology i.e. linear Fresnel concentrator solar power plant.

Finally, we can conclude that, among the above-mentioned sites, the most suitable on performance basis is Tamanrasset followed by Beni-Abbes, then El Oued, and finally Hassi R'mel.

Acknowledgements The authors are grateful to Dr. Said Noureddine from CDER (Renewable Energy Development Center, Algeria) for his constant guidance and critical review in improving the manuscript.

References

1. R.E. Smalley, Future global energy prosperity: the terawatt challenge. MRS Bull. **30**, 412–417 (2005)
2. S. Mihoub, A. Chermiti, H. Beltagy, Methodology of determining the optimum performances of future concentrating solar thermal power plants in Algeria. Energy Int. J. **122**, 801–810 (2017)
3. B. Srimanickam, M.M. Vijayalakshmi, E. Natarajan, Experimental performance assessment of single glazing flat plate solar photovoltaic thermal (PV/T) hybrid system. Prog. Ind. Ecol. Int. J. **9**(2), 111–120 (2015)
4. H. Beltagy, D. Semmar, L. Christophe, N. Said, Theoretical and experimental performance analysis of a Fresnel type solar concentrator. Renew. Energy Int. J. **101**, 782–793 (2017)
5. http://www.Novatecsolar.com
6. ONM/ Office National de la Météorologie, ministère des transports, Algeria
7. http://www.Nrel.Gov/Analysis/Sam
8. H. Beltagy, D. Semmar, N. Said, Performance of medium-power Fresnel Concentrator solar plant in Algerian sites. Energy Procedia **74**(2015), 942–995 (2015)

Chapter 4
Integrating a Solar PV System with a Household Based Backup Generator for Hybrid Swarm Electrification: A Case Study of Nigeria

Rolex Muceka, Tonny Kukeera, Yunus Alokore, Kebir Noara and Sebastian Groh

Abstract Today most of the electrification grids in sub-Saharan Africa (SSA) are found in urban areas. However, these grids experience erratic and frequent power outages for long hours, on average 4.6 h in a day. Due to this problem, many of the African population rely on cheaper but unclean options like backup diesel/petrol generators for lighting, phone charging and other electrical appliances. In Nigeria, millions of people own power generators. These generators are not only noisy but the fuel they use is also costly and result into emissions that pollute the environment. In order to optimize fuel consumption and gradually reduce use of backup generators while increasing share of renewables, a strategy is proposed in this paper to interconnect the existing backup infrastructure to form a bottom-up swarm electrification grid with step by step integration of alternative storages and renewable energy sources. In the swarm-grid excess energy can be generated, sold among grid participants and even at later stage to the national grid. This study

R. Muceka (✉) · T. Kukeera
Pan African University Institute of Water and Energy Sciences (Including Climate Change), B.P. 119, Pôle Chetouane, Tlemcen 13000, Algeria
e-mail: mckrolex@gmail.com

T. Kukeera
e-mail: tonnykukeera@gmail.com

Y. Alokore
Viva Energy International Ltd., P.O. Box 460, Arua, Uganda
e-mail: y.alokore@vivaenergyinternational.com
URL: http://www.vivaenergyinternational.com/

K. Noara
MicroEnergy International, Potsdamer Str. 143, 10783 Berlin, Germany
e-mail: noara.kebir@microenergy-international.com

S. Groh
School of Business and Economics, North South University, Plot 15, Block B, Bashundhara, Dhaka 1229D, Bangladesh
e-mail: sgroh85@gmail.com

M. Mpholo et al. (eds.), *Africa-EU Renewable Energy Research and Innovation Symposium 2018 (RERIS 2018)*, Springer Proceedings in Energy,
https://doi.org/10.1007/978-3-319-93438-9_4

focused on a swarm grid hybrid node consisting of a solar PV system integrated with the existing individual backup generators for households and retail shop end users. The hybrid system designed was found to be a suitable system with fuel savings of 39%, excess energy of 27% and reduced cost of backup electricity by 34% for the household end user. For the retail shop end user, the hybrid system was found to be a suitable system with a fuel cost saving of 53%, excess energy generation of 28% and reduced cost of backup electricity by 45%. The study showed that integration of a solar PV system has a high potential to reduce fuel costs for backup generator end users and presents a great opportunity for hybrid swarm electrification approach.

Keywords National grid · Stand-alone system · Swarm grid · Renewable energy Excess energy

4.1 Introduction

Sub-Saharan Africa (SSA), a region in Africa where most of the population without access to electricity lives. The percentage population in SSA without access to electricity was about 63% in 2014 [1]. The main source of lighting continues to be kerosene lamps, firewood and candles, especially for regions off the main grid. Furthermore, even those with access to the central main grid often suffer from unpredictable power outages for long hours, on average 4.6 h per day, with 17 countries exceeding the average outage duration [2]. According to World Bank's enterprise surveys, last updated in October 2016, the average number of power outages in firms, in a typical month is 8.5 [3]. Many countries experience frequent outages in SSA with the worst scenario seen in Nigeria with 32.8 outages in a month. Other countries with high cases include; the Central African Republic with 29 outages, Congo 21.5, Chad 19.6, Niger 18.5, and Burundi with 16.6 outages [2]. This has led many people in peri-urban areas and trading centres in many countries in Africa to rely on unclean options like backup diesel/petrol generators for lighting, phone charging, and other electrical appliances.

Besides being noisy and producing emissions that are harmful to the environment, backup generators use fuel which makes them costly to maintain and use in the long run. Taking a case for Nigeria; private households spend over $13.35 million USD annually on alternate sources of energy [4]. This figure adds to over $21.8 billion USD per year if enterprises and manufacturers are also considered [4]. For small businesses, fuel costs account for 40% of their total overheads [5].

Owing to a pressing need to protect the environment, cut down fuels costs and promote energy efficiency, as expressed in the sub-goals of the Sustainable Development Goal (SDG) 7 with its main to ensure access to affordable, reliable, sustainable, and modern energy for all, a strategy is proposed to interconnect this existing backup power infrastructure in a swarm grid and to integrate alternative storage and renewable energy generators step by step. By doing so, it is anticipated

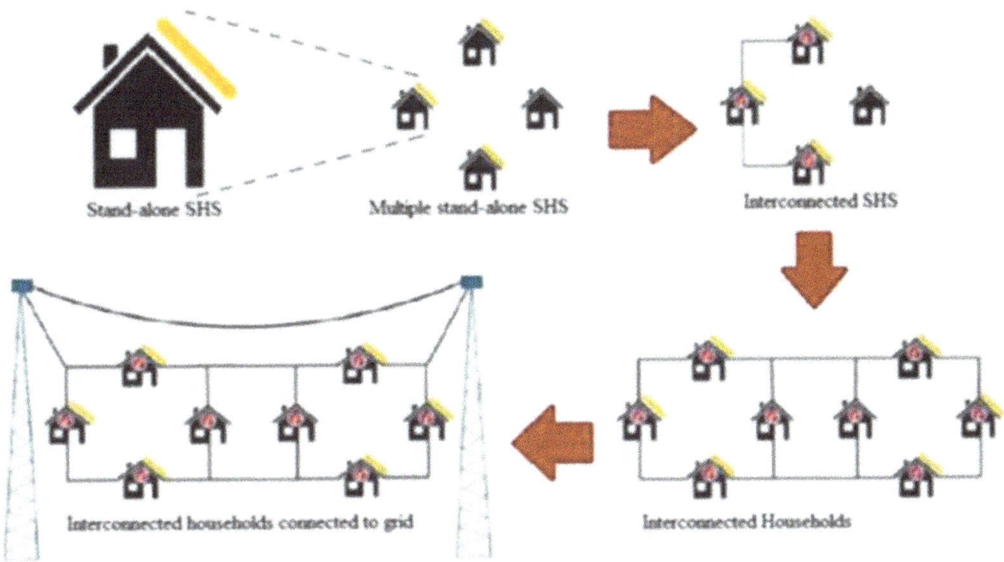

Fig. 4.1 Swarm electrification approach source [6]

that fuel consumption of backup generators could be optimized and gradually reduced, and the share of renewable energy could be increased too. The incremental development of such a grid does not require huge investments from the household or a small enterprise, but can be done using the savings realized on each step. Furthermore, a hypothesis is set up, that each generator in the swarm grid and the swarm grid as a unit can generate excess electricity which can be sold among grid participants and at a later stage even to the national grid. Figure 4.1 shows a representation of swarm electrification approach.

A swarm grid is a grid formed from interconnecting multiple households and small businesses with or without Solar Home Systems (SHS). The concept is known as Swarm Electrification (SE). This concept has been applied in Bangladesh. The end users forming the grid have the ability to share excess energy among themselves. SE allows the households and the small enterprises to become prosumers, i.e. producers and consumers at the same time. As a producer, a node can share or trade the excess energy with its neighbor or as a consumer, a node gets its unmet energy from the neighbors [6]. In a swarm grid a node represents a consumer and/or producer end user.

This paper proposes a strategy of forming a hybrid swarm grid with more than one source of energy in the grid as a whole, and at some nodes in particular. In the following steps, the proposed strategy is described one by one:

1. Integrating an existing generator into a swarm-grid and smart electricity management unit.
    - Excess energy is generated while operating a generator at an optimal point efficiency i.e. 80%.

- The energy can be shared and traded between swarm-grid participants, unlocking capital.

2. Using the capital to invest into batteries, excess energy generated from constantly operating the generators at an optimal point can be stored.
3. The electricity stored in the battery (ies) can be used instead of the generators.

- This reduces the usage hours of a generator hence prolonging its lifespan.

4. Solar PV panels or other renewable energy sources can be integrated to the swarm-grid at the nodes. They can be used as a direct source of power and for charging the batteries.

- Fuel can be saved; excess energy is generated from solar.
- More energy can be shared and traded, unlocking capital for further investment.

5. Step by step, according to the needs, such a grid can grow by adding storage and generation capacities. More users can be integrated and trade with the electricity.
6. Generation and storage capacities might not only be used for backup during power outages but also for extended periods during the day, replacing step by step the power from the grid.
7. When the swarm-grid grows large enough and produces enough excess energy, interconnection with the national grid and a feed in-option can be considered.

This type of grids can grow organically. Each step is voluntary. The economic viability of each step still needs to be proven. The schematic representation of the hybrid swarm grid is shown in Fig. 4.2.

In Nigeria, according to a survey by [7] in the states of Delta, Bayelsa and Rivers, 43% of households with grid connection use backup generators for at least 4 h. Also 40% of the small enterprises with grid connection have less than four hours of electricity supply in a day. In a hybrid swarm grid, a prosumer can be a hybrid node (with more than one energy source at the node) and hence benefits from the merits of combined power sources. A household or small enterprise with a backup fossil-fuel generator could become such a hybrid node by integration of battery and solar PV generator as described in steps 3 and 4 above.

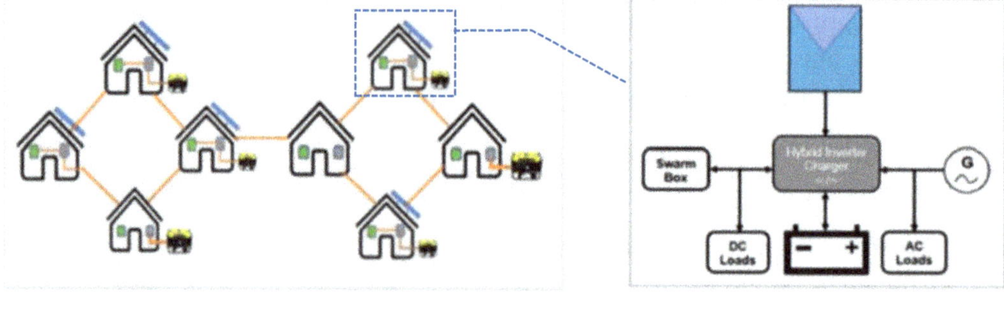

Fig. 4.2 Hybrid swarm grid and a hybrid node

This study focusses on the hybrid node of a swarm grid. With the main objective of contributing to a hybrid swarm grid, the study aimed at designing a hybrid node by integrating a solar PV system for a potential benefit of fossil-fuel savings. The focus was also to quantify excess energy production and determine its potential for sharing in a hybrid swarm grid.

4.2 Methodology

The three states of Bayelsa, Delta and Rivers in Nigeria were chosen as the study areas. The solar energy potential for the study area was determined based on the solar radiation database for PV performance estimation in Europe and Africa [8]. From the literature in [9–11], load profiles were developed for the household and a small enterprise end users as shown in Figs. 4.3 and 4.4.

From the load profiles, the energy demands were estimated, the hybrid units designed controlled by the technical and economic parameters. The energy production and use for the hybrid units designed was further assessed for potential excess energy generation.

4.2.1 Solar Energy Demand

The daily solar energy demand is the solar energy required to charge the battery and as a direct power source in the day [12].

$$W_{pv} = W_{pvbat} + W_{pvday} \tag{4.1}$$

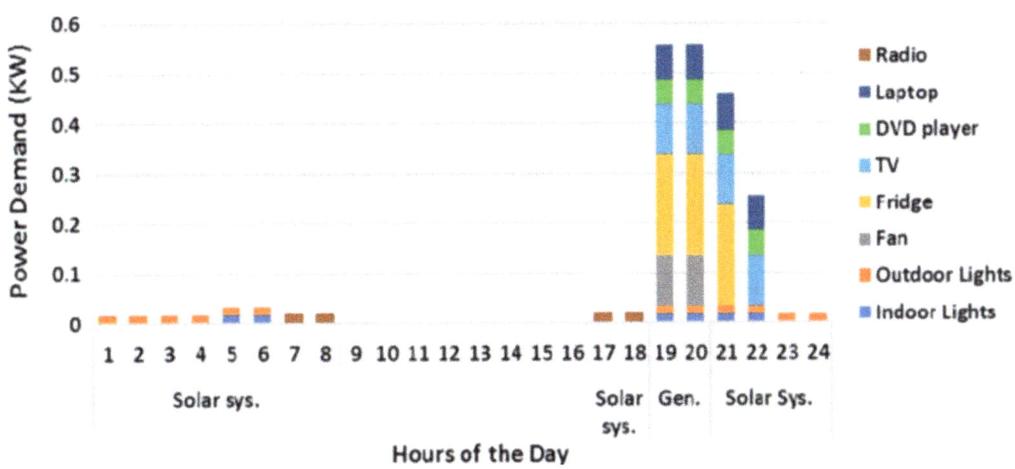

Fig. 4.3 Household end user load profile

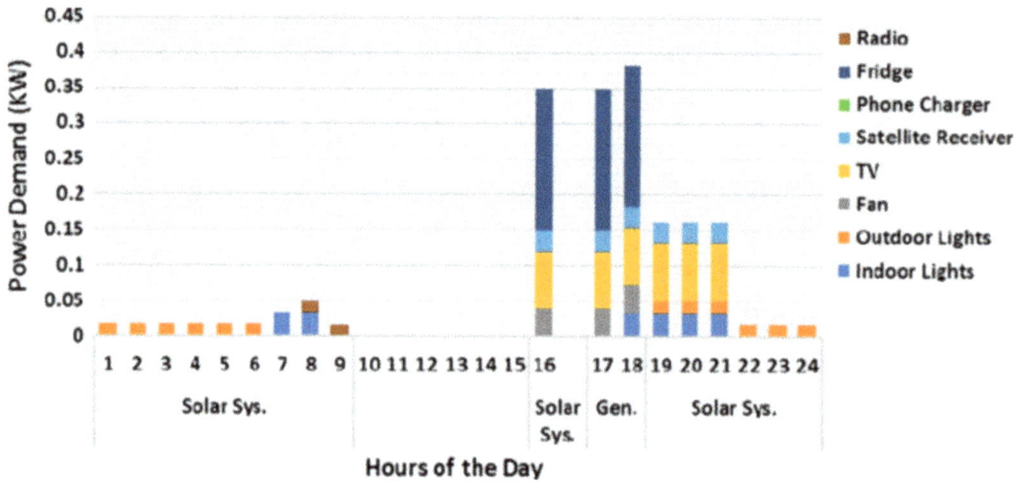

Fig. 4.4 Retail shop end user load profile

where W_{pvbat} is solar energy demand to charge the battery, W_{pvday} is solar energy demand for direct load supply during the day.

4.2.2 *Battery Storage*

The battery is considered like a pot used to store water during the day for later use at night. The method (7) [12] given below is used to size the battery storage;

$$B_B = \frac{E_{bat} \times AD}{V_s \times DOD \times \eta_{inv}} \tag{4.2}$$

where B_B is battery storage size in Ampere hours (Ah), E_{bat} is energy storage demand (Wh), AD is autonomous days, V_s is system voltage, DOD is depth of discharge (60%), η_{inv} is Inverter efficiency (90%).

$$E_{bat} = E_{bat\_night} + (E_d \times 10\%) \tag{4.3}$$

where E_{bat_night} is energy storage demand by loads at night, E_d is day time load energy demand.

In this study, 10% of the day time energy demand is stored to take care of abrupt changes in solar radiation due to swift movements of clouds.

4.2.3 Backup Fossil-Fuel Generator

The rated size of backup generator considered in this study is 1.0 kW each for the household and the small enterprise users, the size is big enough to power the peak demands in the load profiles of the users.

4.2.4 System Solar PV Production

The method applied for the daily estimation of the PV performance is given below [13];

$$P_{pv} = P_{wp} \times \frac{G}{G'} \times \left[1 + \alpha_t\left(T_a - T_{ref}\right)\right] \tag{4.4}$$

where P_{pv} is daily power output of the PV panels at a time, P_{wp} is total PV watt peak under reference conditions, G is daily solar irradiation (W/m^2) at optimal angle (11°), G' is the reference solar radiation (1000 W/m^2); T_a is the daily ambient temperature, and T_{ref} is the reference temperature (25 °C); α_t is the PV panel temperature coefficient; for mono and poly crystalline silicon materials.

4.2.5 Excess/Unmet Energy

Due to the day to day changes in weather, the daily energy production from solar is not the same. On brighter days, the solar irradiations are high and hence high energy production. This energy can be more than what is needed in a day resulting in excess energy and hence is put to waste if not utilized. By performing the solar energy production and consumption analysis, excess/unmet solar energy generation is assessed on daily basis. Excess energy is the energy that could be generated if the battery had not been fully charged or if there was an extra load to be supplied. In this study, the total daily solar energy demand is assumed to be constant in a year.

$$Excess/Unmet_{Energy} = P_{pv} - W_{pv} \tag{4.5}$$

If Eq. (4.5) gives a positive number, it is excess energy and if a negative number, it is unmet energy. During cloudy days or in the rainy season when the unmet energy is most generated, backup fossil-fuel is used to meet the unmet energy.

4.2.6 Economic Analysis

The Annualized Cost of the System (ACS) obtained from Eq. (4.6) is the cost of the system spread or discounted yearly over the whole system lifetime [14].

$$ACS = ACC + AOM + ARC + AFC \tag{4.6}$$

where ACC is the annualized capital cost, AOM is the annualized operation and maintenance cost, ARC is the annualized replacement cost, AFC is the annualized fuel consumption cost.

$$\begin{aligned} ACC &= C_c \times CRF(i', n) \\ CRF &= \frac{i'(1+i')^n}{(1+i')^n - 1} \\ i' &= \frac{i-f}{1+f} \end{aligned} \tag{4.7}$$

where C_c is the capital cost ($ USD), CRF is the capital recovery factor, n is the lifetime of the component in years, f is the inflation rate, i is the nominal interest rate.

$$\begin{aligned} ARC &= C_{rep} \times K - ASV \\ K &= N_{rep} \times SFF(i', n) \\ N_{rep} &= \frac{y}{n} - 1 \text{ if } y \text{ is divisible by } n \\ N_{rep} &= INT\left[\frac{y}{n}\right] \text{ if } y \text{ is not divisible by } n \\ SFF &= \frac{i'}{(1+i')^n - 1} \end{aligned} \tag{4.8}$$

where y is the lifetime of the system, N_{rep} is number of replacement, ASV is annualized salvage value, SFF is sinking fund factor.

$$ASV = S \times SFF(i', y) \tag{4.9}$$

If y is not divisible by n, the salvage value, S of the replaceable component is determined as in [9], with R_l being the remaining life of the component in years.

$$\begin{aligned} S &= C_{rep} \times \frac{R_l}{n} \\ R_l &= n - \left(y - \left(N_{rep} \times n\right)\right) \end{aligned} \tag{4.10}$$

The AOM and the AFC are determined as in [14]

$$
\begin{aligned}
AOM &= \frac{C_c \times (1-\mu)}{n} \\
AFC &= C_f \times f_E \times \sum_{t=1}^{365} E_{gen}
\end{aligned} \tag{4.11}
$$

where C_c is the capital cost, μ is the reliability of the component and n is the lifetime of the component.C_f is the fuel cost per litre in \$ USD/l, f_E is the fuel consumption per unit energy (l/kWh), E_{gen} is the backup fossil-fuel generator daily energy output (kWh).

4.2.7 *Levelized Cost of Backup Electricity (LCoE)*

The cost of energy paid for the electricity produced and used in a year is computed using Eq. (4.12) where E_{year} is the electrical energy consumed in a year.

$$
LCoE = \frac{ACS}{E_{year}} \tag{4.12}
$$

4.3 Results and Discussion

A hybrid node is designed each for a household and a small enterprise. The results are discussed below in comparison to the baseline system of having a fossil-fuel backup generator.

4.3.1 *Technical and Economic Analysis*

The technical and economic results for the designed hybrid systems are as shown. From Table 4.1, the baseline system consists of the fossil-fuel backup generator as the main power source during power outage. The designed system is the hybrid system to be at a node of a swarm grid consisting of both a fossil-fuel backup generator and the PV system generator.

From Table 4.2, the designed hybrid systems are more cost effective than the baseline system by the end of their life time, as shown by lower values of the annualized costs of the systems. The levelized cost per unit of energy used is also lower for the designed system. The system costs are less due to reduced fuel consumption costs. The fuel costs reduced because of reduced usage hours of the fossil-fuel generators as more solar energy is used.

Table 4.1 Technical analysis for the end user systems

End users	System components	Generator (kW)	Battery (Ah)	Inverter (VA)	Solar PV (Wp)
Household	Baseline system	1.0	–	–	–
	Designed system	1.0	150	850	300
Retail shop	Baseline system	1.0	–	–	–
	Designed system	1.0	200	500	400

Table 4.2 Economic analysis for the end user systems

End users	Economic parameters	AFC ($ USD)	LCoE ($ USD/ kWh)	ACS ($ USD)	ICS ($ USD)
Household	Baseline system	349.12	0.574	381.89	137.3
	Designed system	213.48	0.447	357.81	932.1
Retail shop	Baseline system	299.74	0.582	332.51	137.3
	Designed system	140.96	0.403	320.01	1069.6

However, the Initial Cost of System (ICS) is higher for the designed hybrid systems as compared to the baseline systems. This could highly prevent the end users from integrating solar PV generator systems. Nevertheless, inclusive and flexible financial solutions could be used to make the systems more affordable as payment would only apply for the integrated solar PV system since the fossil-fuel generator as the baseline is already owned by the end users.

4.3.2 Energy Analysis

The production and consumption of different energy sources for the hybrid systems are analysed. Solar PV production and use in particular is also analysed. Excess energy potential is assessed and it is on that basis that recommendation for energy sharing in swarm grid is made.

From Fig. 4.5, fossil-fuel generator is still slightly used more throughout the year than the integrated solar energy for the household end user, however, its use has reduced to 51% with 49% integration of solar PV system as indicated in Fig. 4.6.

Figure 4.7 shows solar energy production is high at beginning and towards the end of the year. With the demand maintained constant, total excess solar energy of 27% is found for the household as illustrated in Fig. 4.8. The excess energy is

Fig. 4.5 Household end user energy consumption analysis

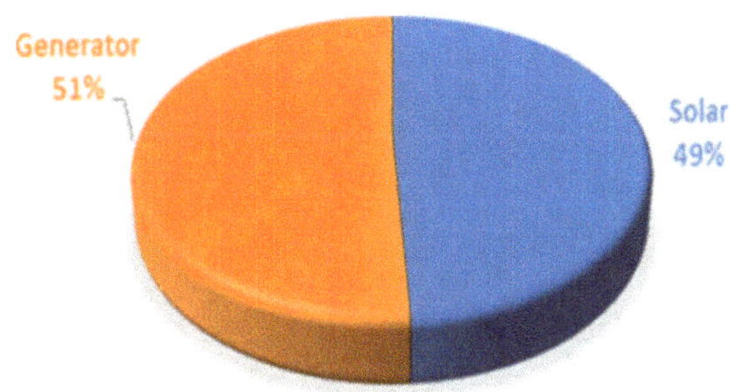

Fig. 4.6 Household energy mix

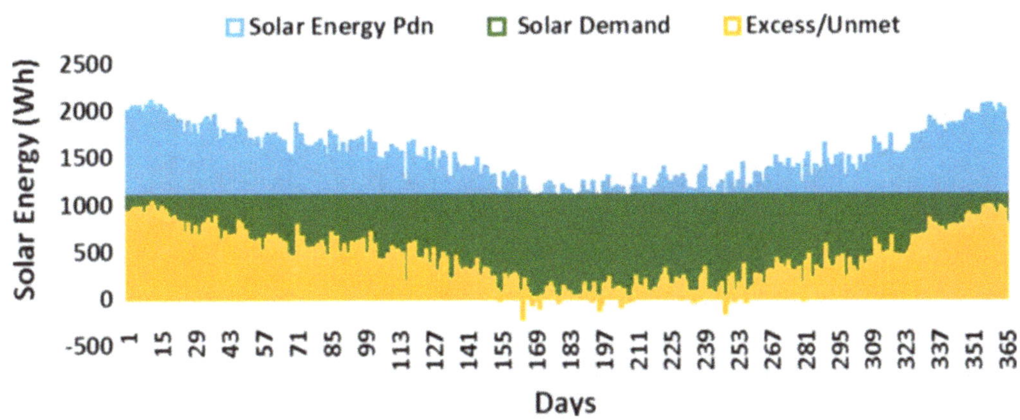

Fig. 4.7 Household end user solar energy production analysis

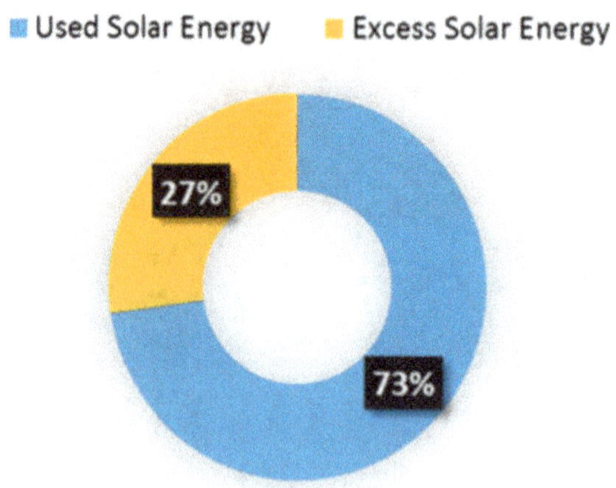

Fig. 4.8 Household end user ratio of excess solar energy

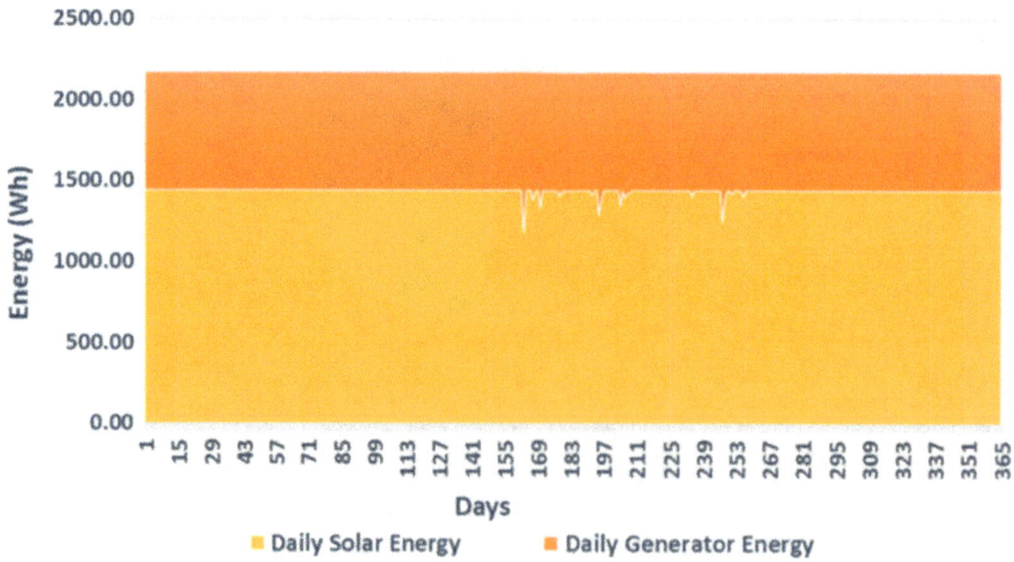

Fig. 4.9 Retail shop end user energy consumption analysis

generated on daily bases except in the middle of the year during rainy season when unmet energy is realized. The unmet energy is then supplied by the fossil-fuel generator as shown in Fig. 4.5.

For the retail shop enterprise, the integrated solar energy PV system becomes the main source throughout the year. Solar energy accounts for 66% of the daily use and fossil-fuel generator, 34%. See Figs. 4.9 and 4.10.

Figure 4.11 for the retail shop end user shows solar energy production is high at beginning and towards the end of the year similar to the household end user as they are of the same geographical location. The fossil-fuel generator is used to meet the unmet solar energy in the rainy days especially in the middle of the year, see Figs. 4.9 and 4.11. Like for the household, the excess energy for the retail shop end user is found to be 28% (Fig. 4.12) and is also generated on daily bases.

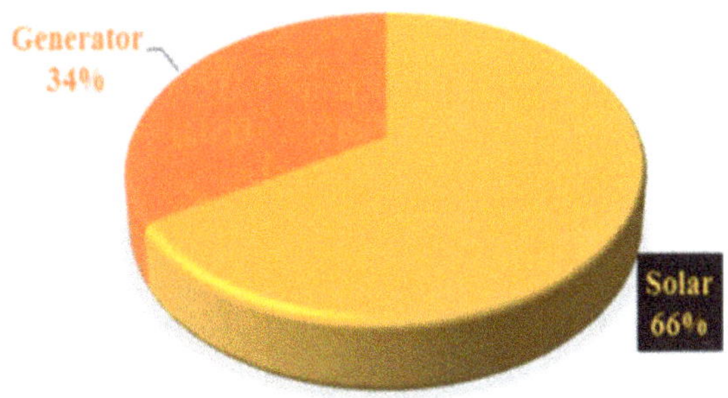

Fig. 4.10 Retail shop end user energy mix

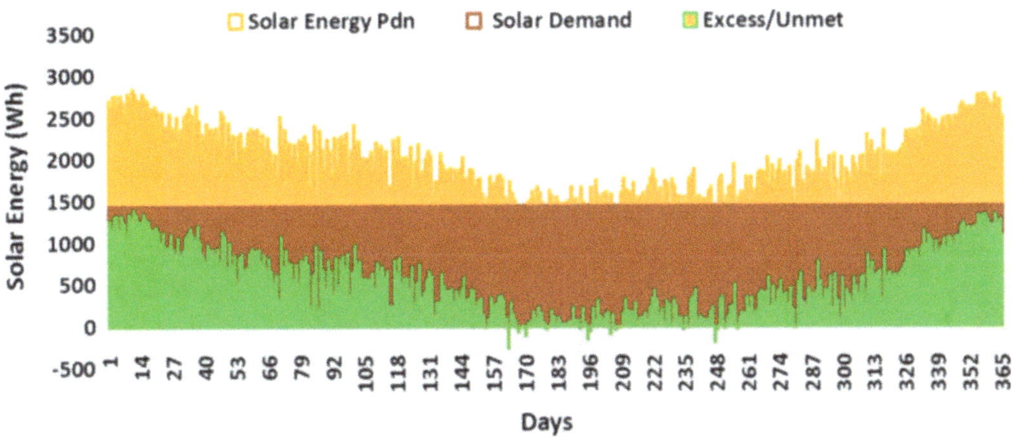

Fig. 4.11 Retail shop end user solar energy production analysis

Fig. 4.12 Retail shop end user ratio of excess solar energy

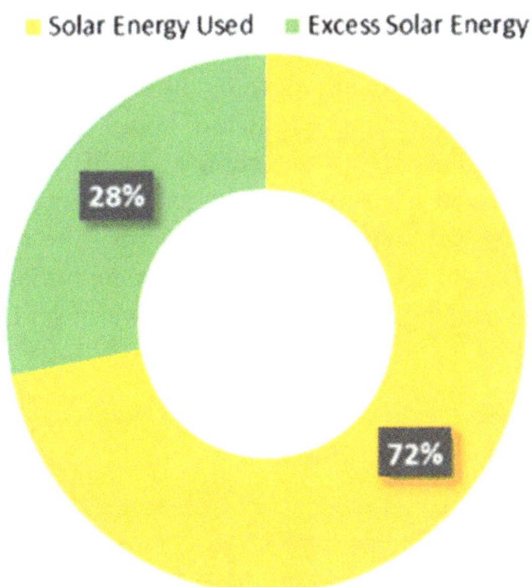

Table 4.3 Summary of results

End users	Systems	LCoE ($ USD/kWh)	Reduction in LCoE (%)	AFC ($ USD)	Reduction in AFC (%)	Excess energy (Solar) (%)
Household	Baseline	0.574	–	349.12		
	Designed system	0.447	22	213.48	39	27
	Designed system (Swarm grid)	0.378	34			
Retail shop	Baseline	0.582	–	299.74		
	Designed system	0.403	31	140.96	53	28
	Designed system (Swarm grid)	0.321	45			

4.3.3 Hybrid Swarm Grid Assessment

The result of excess solar energy found in this study is similar to that of Kirchhoff, in which he found an excess energy of 30% of the potential solar energy production from a single solar home system [15], a typical case of Bangladesh. In a swarm grid the excess energy can be utilized instead of being wasted. Also, the excess energy can be utilized by connecting more loads when swarm grid infrastructure is still missing.

In a swarm grid where excess solar energy can be traded, Table 4.3 shows that the LCoE could reduce by 34% for the household and 45% for the retail shop end users operating as hybrid nodes in a hybrid swarm grid. In comparison to Bangladesh, the reduction in LCoE in this study is higher. This is a result of reduced use of fossil fuel generators while the systems in Bangladesh are purely single source solar home systems. The reduction in LCoE would enable the end users to access reliable and a more affordable electricity while supporting neighbors by trading excess energy in a swarm grid.

4.4 Conclusion

Power outages in SSA occur at a high rate, the situation is more alarming in Nigeria. This has resulted into high reliance on fossil-fuel backup generators for most households and small enterprises. However, these generators are noisy and their fuel is costly. This study looked at a stepwise strategy of integrating solar PV system with a backup household-based generator interconnected in a swarm grid for a potential benefit of fuel savings and environmental protection in line with SDG 7. The study focussed on the hybrid node of the swarm grid. From the result, a potential fuel savings and overall system cost reduction is found. Integrating solar, hence increases the share of renewable energy mix for the end users, generates excess energy that can be shared in a swarm grid by the grid participants. The overall result is that swarm grid would enable the energy consumers to become prosumers capable of sharing or trading energy produced among themselves. However, for a successful implementation of this concept in SSA, the social-cultural aspects of sharing electricity in a swarm grid should be investigated.

References

1. IEA and World Bank, *Sustainable Energy For All Global Tracking Framework Progress Toward Sustainable Energy* (2017)
2. IRENA, *Solar PV in Africa: Costs and Markets* (2016)
3. World Bank, Power outages in firms in a typical month (number). *Enterprise Surveys* (2016)
4. D.J. Ley, K. Gaines, A. Ghatikar, *The Nigerian Energy Sector—An Overview with a Special Emphasis on Renewable Energy, Energy Efficiency and Rural Electrification* (2014)
5. The Economist Intelligence Unit Limited, *Enabling a More Productive Nigeria: Powering SMEs* (2015)
6. S. Groh, D. Phillip, B. Edlefsen Lasch, H. Kirchhoff, Swarm electrification—investigating a paradigm sift through the building of microgrids bottom-up, in *Decentralized Solutions for Developing Economies*, eds. by S. Groh, J. Van der Straeten, B. Edlefsen Lasch, D. Gershenson, W. Leal Filho, D. Kammen. Springer Proceedings in Energy XXIV (2015), pp. 3–22
7. All On and Shell, *Nigeria: Energy Needs Assessment and Value Chain Analysis* (2016)
8. European Commission, Photovoltaic Geographical Information System (PVGIS) (2017) [Online]. http://re.jrc.ec.europa.eu/pvg_tools/en/tools.html#HR. Accessed 04 June 2017
9. V. Anayochukwu Ani, Design of a reliable hybrid (PV/diesel) power system with energy storage in batteries for remote residential home. *J. Energy* **2016** (2016)
10. V. Anayochukwu Ani, Feasibility analysis and simulation of a stand-alone photovoltaic energy system for electricity generation and environmental sustainability—equivalent to 650VA fuel-powered generator—popularly known as 'I pass my neighbour'. *Front. Energy Res.* **3**(Sept), 1–9 (2015)
11. O.S. Omogoye, A.B. Ogundare, I.O. Akanji, Development of a cost-effective solar/diesel independent power plant for a remote station. **2015** (2015)
12. P. Mohanty, T. Muneer, Smart design of stand-alone solar PV system for off grid electrification projects, in *Mini-Grids for Rural Electrification of Developing Countries*, eds. by S.C. Bhattacharyya, D. Palit. Green Energy and Technology (Springer, Cham, 2014), pp. 63–94
13. P.J. Boait, Technical aspects of mini-grids for rural electrification, in *Mini-Grids for Rural Electrification of Developing Countries*, eds. by S.C. Bhattacharyya, D. Palit. Green Energy and Technology (Springer, Cham, 2014), pp. 37–61
14. H. Suryoatmojo, A.A. Elbaset, F.A. Pamuji, D.C. Riawan, M. Abdillah, Optimal sizing and control strategy of hybrid PV-diesel-battery systems for isolated Island*. no. 1, 1–6
15. H. Kirchhoff, *Microgrid Concepts with an Agent-Based Control Scheme in the Context of the Electrification in Off-Grid Areas* (Technical University of Berlin, 2013)

Chapter 5
Overview of Economic Viability and Social Impact of Renewable Energy Deployment in Africa

A. Khellaf

Abstract Africa is endowed with a very important renewable energy potential. Exploited, this potential permits the continent not only to get out of the energy poverty it is suffering from but also to ensure a sustainable development. In the present work, the economic viability of renewable power generation is assessed and analyzed. Then an identification and discussion of the impacts that the renewable energy deployment could have on socio-economic progress, health, education, gender equity and rural development are carried out.

Keywords Africa · Electricity access · Renewable energy · Economic viability Social impacts · Health impacts

5.1 Introduction

Africa suffers from energy deficit. Indeed, though it represents 16% of the world population, it consumes only 3% of the power generated worldwide. More than 600 million Africans have no access to electricity [1]. The electrical networks are also plagued by technical problems. Because of aging and poorly maintained transmission and distribution networks, the electrical grids experience big losses and recurrent outage. In fact, 95% of the African countries experience electrical losses larger than the world average losses. Countries, such as Libya and Togo, experience staggering losses respectively of the order of 69.7 and 72.5% [1].

The grid system suffers also from high recurrence and high duration of outages. It has been reported that Senegal, Tanzania and Burundi experience unbelievable outages that sum up respectively to 45, 63 and 144 days/year [2]. Moreover, Africa suffers from a lack of access to clean fuels and modern technology for cooking.

A. Khellaf (✉)
Centre de Développement des Energies Renouvelables,
BP 62, Route de l'Observatoire, 16340 Bouzareah, Algiers, Algeria
e-mail: a.khellaf@cder.dz

M. Mpholo et al. (eds.), *Africa-EU Renewable Energy Research and Innovation Symposium 2018 (RERIS 2018)*, Springer Proceedings in Energy,
https://doi.org/10.1007/978-3-319-93438-9_5

About 60% of the African population rely on traditional firewood or charcoal air polluting stoves for cooking. The damage to health is enormous [3].

The present work deals with the deployment of renewable energy. More particularly, an assessment and an analysis of the viability of renewable energy deployment is carried out. The impacts of this deployment on socio-economic progress, health, education, gender equity and rural development are then identified and discussed.

5.2 Economics of Renewable Energy Deployment

The capacity of renewable energy deployment depends on a large part on the initial financial investment that is required for renewable power generation infrastructures deployment. The economic success of the renewable energy deployment is related to its financial viability, which depends on the deployment capacity in generating energy at a competitive cost.

5.2.1 Capital Investment Cost

Renewable energy deployment is capital investment intensive. In Table 5.1, capital costs of renewable and non-renewable power generation deployment are reported [4, 5]. From this table, it can be seen that higher capital costs are required for

Table 5.1 Capital cost for renewable and non-renewable power generation systems [4, 5]

Source	Technology	Installation cost ($/W)
Conventional source	Nuclear plant	4.27–7.93
	Natural gas combustion turbine	0.49–0.81
	Natural gas combined cycle	0.92–1.54
	Coal fired power plant	1.88–3.90
Renewable source	Stand-alone biomass power plant	2.87–5.75
	Geothermal power plant	5.94
	Hydropower	2.28–4.73
	Solar PV power plant	2.12–3.54
	Solar PV home system	3.6–17.0
	Grid connected solar PV roof top system	2.0–3.0
	Mini-grid system	2.5–2.9
	CSP trough with storage power plant	4.59–8.12
	CSP tower with storage power plant	4.58–8.10
	Wind on shore power plant	1.49–2.48
	Wind fixed bottom off shore power plant	2.15–4.47
	Wind floating platform off shore power plant	2.73–5.67

renewable energy deployment than for non-renewable deployment. Nonetheless, it is worth to note that wind technology deployment is competitive with the non-renewable source technologies. It has also been argued that, as the conventional source technologies are mature, it is more likely that their capital cost will not change much over the years. On the other hand, renewable technologies, with improvement in technologies and with the learning curve effect, capital costs are undoubtedly going to drop; making them more competitive with conventional energy source technologies.

Given the actual state of most of the African states economies, capital cost could be a serious hurdle for most of them. Fortunately, there are many funding opportunities focused or related to renewable energy that must be grabbed.

These opportunities should be sought from local or public sectors or from the development finance institutions. Among the development finance institutions, there is the World Bank that is ready for renewable energy deployment through its Climate Investment Fund and its Strategic Climate Fund [6]. There is also the African Development Bank which gets involved in renewable energy deployment through its Sustainable Fund for Africa and its Partial Risk Guarantees. Green Facility for Africa, African Development Fund, International Climate Fund, Global Environment Fund, Global Energy Efficiency and Renewable Energy Fund and Global Environment Facility Trust Fund are development finance institutions that are deeply involved in renewable energy deployment in Africa. Funding in Africa could be under the form of a loan, a rebate, a subsidy or a grant. The funding could also take the form of a technical support or an action of awareness and training.

5.2.2 Renewable Power Generation Economic Viability

There are different methods for assessing the economic viability of renewable energy production facilities deployment. The assessment could, for example, be based on the net revenue or on the production cost relative to the cost of production of a non-renewable energy production technology [3, 7]. The levelized cost of energy (LCOE) is usually used to provide an indication of the competiveness of energy generation of a given technology. In Table 5.2, the levelized cost of power generation based on renewable and non-renewable energy sources are reported [8, 9]. From this Table, it can be noticed that Solar PV and wind systems connected to the grid are competitive with gas turbines. The same way, geothermal, bio-power and hydropower connected to the grid are competitive with gas turbines and combined cycles. Mini-grid and off-grid technologies for solar PV and wind remain expensive by comparison to grid connected technologies. Nonetheless, they offer flexibility and if power transport is taken into consideration, they will be more attractive. It must be added that different local aspects could affect the economic viability of renewable energy deployment:

1. The conversion technology in association with the local technical potential of the renewable energy source;

Table 5.2 Levelized cost of energy for power generation [8, 9]

Source	Technology	LCOE ($/kWh)
Conventional source	Gas turbine	0.10–0.15
	Combined cycle	0.055–0.10
	Coal	0.045–0.05
Renewable source	Solar PV utility	0.06–0.26
	CSP power plant	0.28–0.33
	Wind utility	0.05–0.17
	Geothermal	0.045–0.13
	Bio-power	0.04–0.18
	Hydropower	0.04–0.2
	Off- or mini-grid solar PV	0.30–0.32
	Off- or mini-grid wind	0.25–0.26
	Small hydro	0.25–0.28

2. The local cost of material, labor and other matters that vary locally;
3. The location of the power system in relation to the transmission lines and to the consumers.

For wind energy, it has been reported that power generation is viable for wind speeds larger than 4 m/s at 80 m above ground level. For solar PV, power generation is viable in the case where the global horizontal irradiance is larger than 1000 kWh/m^2 year; while concentrated solar power (CSP) generation is viable in the case where direct normal irradiance (DNI) is above 1800 kWh/m^2 year. It has also been found that the levelized cost of energy for power generation using CSP decreases with increasing DNI [10].

5.3 Renewable Energy Deployment Social Impacts

Renewable energy deployment has social impacts on different levels. The impacts could be at the socio-economic level, the improvement of quality of life, health and education.

5.3.1 Socio-Economic Impact

Renewable energy influences positively the contribution to industrial development and job creation throughout the whole chain of the energy deployment. As shown in

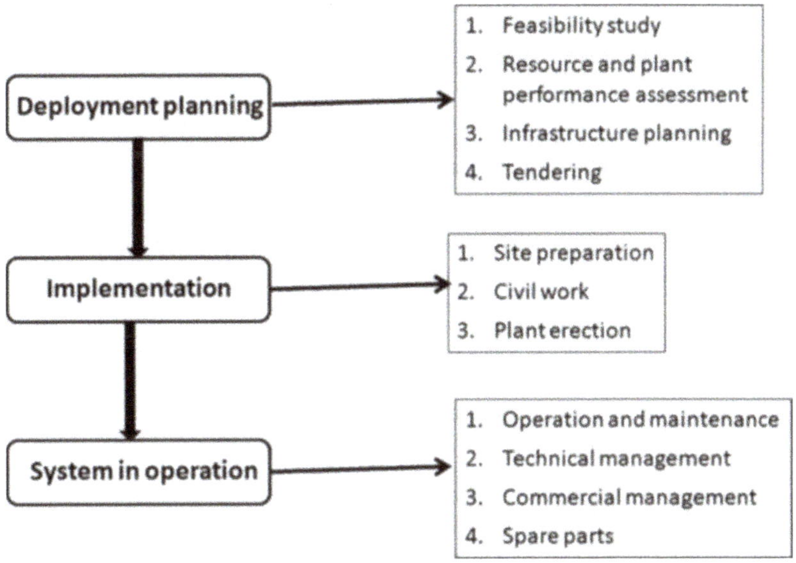

Fig. 5.1 Chain of energy project deployment

Fig. 5.1, the impacts occur at different phases of the deployment: from the deployment-planning phase to the system operation phase [11].

During these phases, the involvements of the local experts, consultants and specialized firms are necessary. In the deployment-planning phase, the involvements are in the feasibility studies, the resource and plant performance assessment, the infrastructure planning and the tendering. In the implementation phase, the involvements are in the site preparation, the civil work and eventually in the plant erection. Finally, for the system operation, the involvements are in the actions of operation and maintenance, the technical management, the commercial activities and the renewable energy parts manufacturing.

For the local involvements to be successful, there should be sound supporting processes. These processes include [11]:

- Policymaking: local policy makers and administrations play, through deployment strategies and policies, a central role in ensuring an effective deployment of the renewable energy.
- Finance: finance services are important as they play a major role not only in renewable energy investment promotion but also in attracting international investors and finance management.
- Education and capacity building: these two actions are necessary for the supply of the local qualified labor and for the creation of innovative businesses and firms. It is also the best way of ensuring a sound technology transfer.
- Research and development: It is by all means the way to ensure growth and innovation. Indeed, it is through research and development that innovative technologies could be implemented. In the long term, it ensures efficient, sound and economically viable energy systems.

- Creation of small and medium enterprises (SME) and small and medium industries (SMI): the establishment of these SME/SMI ensures not only creation of jobs but also economic growth. It ensures also the availability of service entities that insure, among others, maintenance and spare parts.
- Renewable energy and related industry: the development of a local industrial tissue is essential for an efficient and economically viable renewable energy deployment. This industry could include raw materials as well as component production and assembly.

5.3.2 Health Impacts

An analysis of the African primary energy use shows that the use of traditional biomass fuel, constituted mainly of firewood and charcoal, outweighs the demand for all other forms of energy combined. The share of biomass fuel in the primary energy mix is more than 84% in the residential sector and about 7% in the industrial sector [12].

Figure 5.2 shows the percentage of the population that rely on biomass bio-fuel. It can be noted from this Figure that 60% of the African population, i.e. about 730 million, rely on polluting fuel. In the case of Benin, Burkina Fasso and more particularly Uganda, this percentage is very high; it is more than 94% [12]. It should also noted that almost 850 million have no access to clean cooking.

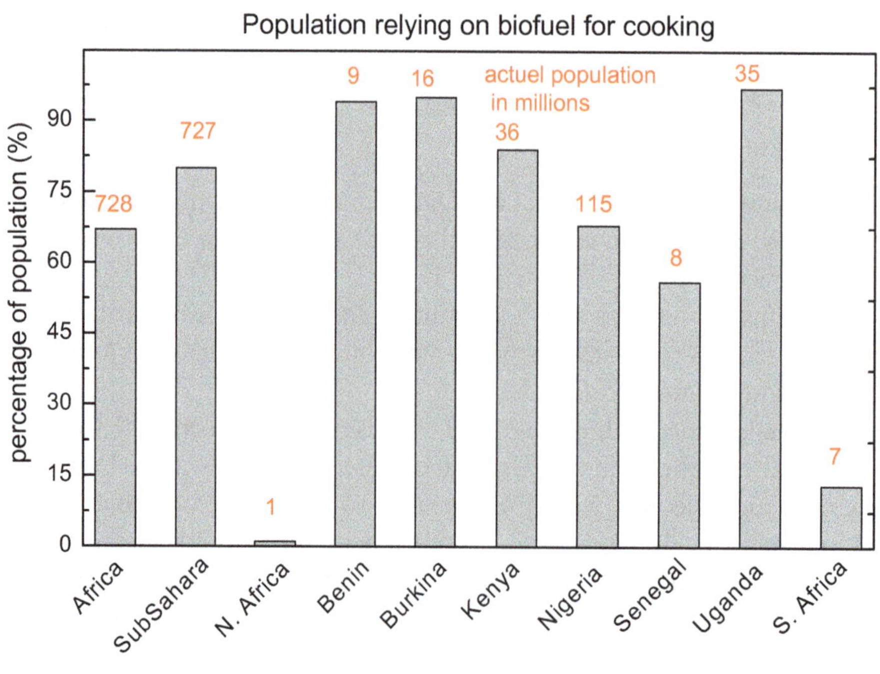

Fig. 5.2 Portion of the African population relying on biofuel [12]

The traditional use of biofuel not only causes pollution and leads to deforestation but is also a major source of health problems. It has been reported that the use of firewood or charcoal stoves are the reason for the spread of very serious illnesses such as respiratory infections, strokes, lung cancer and pulmonary diseases. These low efficiency stoves expose people living in particularly poorly ventilated houses to carbon monoxide and other toxic matters. Women and infants, who spend most of the time indoor, suffer the most.

The deployment of renewable energy introduces affordable, clean and modern energy technologies. This reduces and eventually phases out the need for traditional biomass fuel. This undoubtedly facilitates the introduction of clean energy and modern technology stoves. This leads to a drastic reduction of indoor pollution, reducing and eventually eliminating the health threat associated with this phenomenon. Infants and women wellbeing would then increase.

Moreover, the deployment of renewable energy can improve people's health by enhancing health services. Survey by the World Health Organization has found that actually 58% of health institutions in a large number of African countries have no access to electricity [12]. Indeed, by assuring the electrification of the health institution infrastructures, it is possible to offer modern health services such as the use of modern health equipment for diagnostic and treatment [13]. Electricity is needed for refrigeration of pharmaceutical products, medical tools sterilization and for emergency procedures.

5.3.3 Impacts on Education

Actually, a large number of African primary schools are without electricity. It has been falsely argued that schooling takes place during the day. This argument has led to many African countries, as shown in Fig. 5.3, neglecting school electrification. As shown from this Figure, only five countries, namely, Djibouti, Guinea-Bissau, Namibia, Rwanda and Swaziland reported primary school electrification rate higher than the national electrification rate. For at least 12.5% of the African countries, the primary school electrification rate is less than 5%. Central African Republic holds the grim record of no primary school electrification. It should be noted that Seychelles and Mauritius have full primary school electrification, while Algeria has a 97% public school electrification [14].

Arguing that schooling takes place during daytime in order to delay school electrification is false and harmful. It is false as schooling could be extended to after daytime and that homework and class preparation usually take place at night. It is also false in the sense that electricity is needed not only for lighting but also for the energy that powers equipments, such as computers and laboratory instrumentations that are necessary for the education. It is harmful in the sense that the process that ensure the provision of a quality education is overlooked. Indeed, a lack of electrification usually leads not only to pupils' failure with a large drop out but this could hinder the attraction of the most competent teachers and their retention.

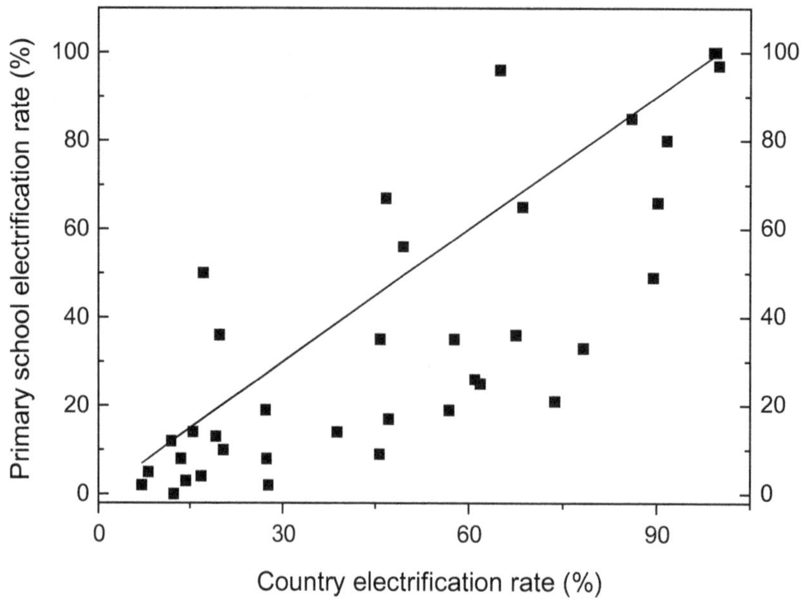

Fig. 5.3 Primary school electrification rate as function of the country electrification rate for African countries (source data [1, 14])

Renewable energy deployment, by its flexibility and adaptability to different regions, is capable of ensuring a viable schools electrification. Among the positive impacts of the renewable, there is [15].

(a) Light and study time: In most non-electrified African schools, school attendance is limited to daytime. Moreover, in this case, homework and course preparation are usually carried out using inefficient, polluting and health threatening kerosene lamps. Electrification enables the provision of clean lightening. Studies have shown that the electricity light is cleaner and economically more viable [16]. Moreover, the electricity light allows an increase in the use of the school infrastructure. This allows the extension of teaching time to early morning and to late afternoon [17]. School infrastructures could also be used by pupils to do homework and by teachers to prepare courses. Studies have shown that school electrification increases youth literacy rate [18].

(b) School performance: Undoubtedly, school electrification has positive impact on school performance, leading to less absenteeism, more enrolments and larger numbers of graduation. Studies carried out in Sudan, Tanzania and Kenya [19] have shown that school electrification result in doubling the number of pupils completing their academic curriculum. It has been argued that providing electricity leads to a much better academic performance [20].

(c) ICT as education tools: Information and communications technologies (ICT) offer many opportunities for high quality education experiences, both in-class as well as through e-learning. Experiences in using renewable energy for ICT introduction in schools have been successfully carried out [21]. Renewable energy deployment holds promises for a strong ICT involvement in education.

The ICT technologies include telephone, radio, television, audio and video tapes. These technologies have enhanced the pupils' ability to learn and improve their academic achievements [22]. But it is with the advent of computer and internet that ICT became potent central tools for education and information gathering [23]. In Africa, Rwanda and Namibia have taken steps for ICT introduction in Education [22].

(d) Teacher attraction and retention: Best teachers usually choose schools that offer them the best chance to develop their skills and to evolve by acquiring new ones. It is no surprise that school electrification is at the top of their choice [24]. Indeed, electrification enables them to acquire the tools, such as ICT, which facilitate their teaching and allow them to progress in their career.

(e) Gender enrolment: School electrification has increased girls' enrolment in education. Studies have shown that there is a direct correlation between school electrification and the ratio of girl-to-boy enrolment in school [25]. School electrification in the mostly nomad populated area of southern Algeria has been the best way of convincing the nomad population to enrol their children in boarding schools [26].

5.3.4 Impacts on Rural Development

Rural area dwellers lag urban dwellers in access to electricity and to clean fuels and technologies for cooking. Figure 5.4 reports the electrification rates for Africa as a

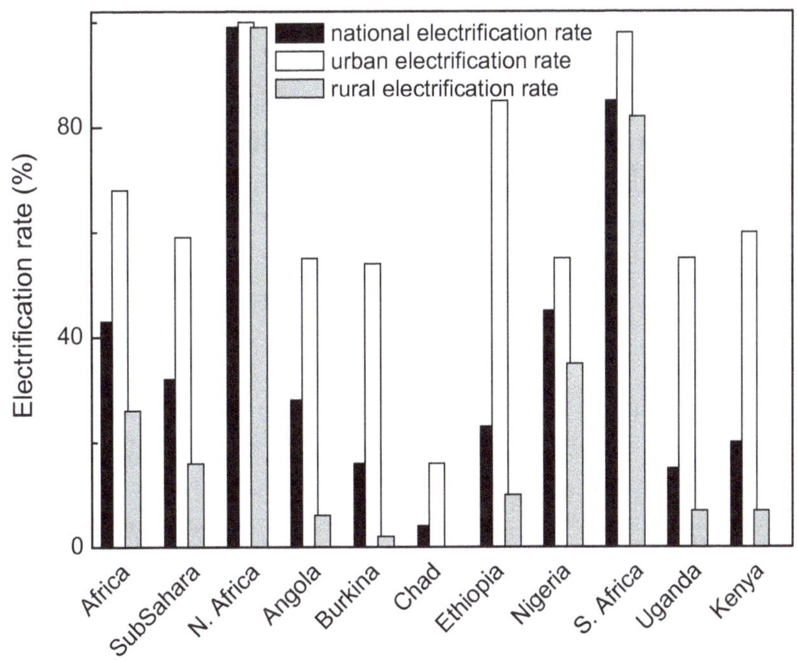

whole and for different African countries. From this Figure, it can be noted that at the African level, the urban electrification rate is 161% more important than the rural one. For some countries, such as Angola, Burkina Faso and Kenya, the rural electrification rate is extremely low. For these countries, the urban electrification rates are respectively 817, 2600 and 757% more important than the rural electrification rates. It has been reported that about 480 million rural Africans have no access to electricity [27].

This could be explained in part by the power grid system adopted by the African countries. This system is centralized and as such is not well adapted to scarcely populated areas such as the Sahara and the Kalahari deserts or to scattered dwellings such as the case of the African rural areas. It should be noted that 60% of the African population live in such areas. The extension of the centralized grid to these scarcely populated or scattered dwellings areas is financially prohibitive and technically challenging.

Nonetheless, the off-grid technologies, using renewable energy sources, have proven to be appropriate for the electrification of the rural areas. These technologies, either on stand-alone, micro-grid or mini-grid, have become the options for localities where central grid connections are neither available nor affordable.

The first impact is an improvement in human health and a reduction of indoor pollution, as a result of replacing kerosene lamps by electricity for lighting and firewood stoves by modern stoves for cooking.

With electrification, there are possibilities of using household appliances, such as refrigerators and blenders. This has the effect of not only improving people's diet but also saving time that could be used in more productive activities.

Electrification offers also the opportunities for local education. Studies have shown that rural electrification has led to an increase in rural school enrolment [28] and to a reduction in rural exodus [29].

Electrification can also boost agriculture productivity. By allowing irrigation through water pumping and by permitting crops storages and processing, electrification offer the opportunity to the farmers to contribute to the agro-industry sector and by that increase their incomes.

5.4 Conclusion

Africa's energy deficits and technical power problems have been hampering its sustainable economic growth. To solve these problems, the African countries are resorting to the development of their important renewable energy potential. National energy programs have been devised and policy instruments for their implementations identified.

An analysis of the renewable energy deployment economy shows that wind energy and solar PV are viable and competitive with conventional energy. Moreover, renewable energy deployment impacts positively on socio-economic progress, health, education, gender equity and rural development.

References

1. Worldbank datadbank. http://databank.worldbank.org/data/databases.aspx
2. A. Eberhard, O. Rosnes, M. Shkaratan, H. Vennemo, in *Africa's Power Infrastructure: Investment, Integration, Efficiency* (The World Bank, Washington DC, 2011)
3. IEA, Energy Access Outlook 2017, World Energy Outlook Special Report (2017)
4. Black & Veatch, in *Cost and Performance Data for Power Generation Technologies*. Report prepared for the National Renewable Energy Laboratory (2012). http://bv.com/docs/reports-studies/nrel-cost-report.pdf
5. Energy and Environmental Economics. *Capital Cost Review of Power Generation Technologies: Recommendations for WECC's 10- and 20-Year Studies* (San Francisco, CA, 2014)
6. G. Schwerhoff, M. Sy, Financing renewable energy in Africa—key challenge of the sustainable development goals. Renew. Sustain. Energy Rev. **75**, 393–401 (2017)
7. A. Brown, P. Beiter, D. Heimiller, C. Davidson, P. Denholm, J. Melius, A. Lopez, D. Hettinger, D. Mulcahy, G. Porro, *Estimating Renewable Energy Economic Potential in the United States: Methodology and Initial Results*, Technical Report NREL/TP-6A20-64503 National Renewable Energy Laboratory Revised August 2016
8. US-IEA, Levelized Cost and Levelized Avoided Cost of New Generation Resources, Annual Energy Outlook 2017
9. REN21, Renewables 2016, Global Status report. www.ren21.net
10. Irena, *Renewable Energy Technologies Cost Analysis Series, Volume 1: Power Sector*, Concentrating Solar Power, June 2012
11. IRENA, The socio-economic benefits of solar and wind energy (2014). www.irena.org/Publications
12. J. Lee, 5 Reasons to care about access to electricity, United Nations Foundation (August 2013). http://www.unfoundation.org/blog/5-reasons-electricity.html
13. UNESCO Institute for Statistics, A view inside schools in Africa: regional education survey (2014)
14. UNDESA, Energy and Education 1: electricity and education: the benefits, barriers, and recommendations for achieving the electrification of primary and secondary schools (December 2014)
15. E. Adkins, S. Eapen, F. Kaluwile, G. Nair, V. Modi, Off-grid energy services for the poor: introducing LED lighting in the Millennium Villages Project in Malawi. Energy Policy **38**, 1087–1097 (2010)
16. C. Kirubi, A. Jacobson, D. Kammen, A. Mills, Community-based electric micro-grids can contribute to rural development: evidence from Kenya. World Dev. **37**(7), 1208–1221 (2009)
17. A.S.A.C. Diniz, E.D. França, C.F. Câmara, P.M.R. Morais, L. Vilhena, The important contribution of photovoltaics in a rural school electrification program, in *Transactions of the IEEE* (2006), pp. 2528–2531
18. Phil Goodwin, The dark side of education (October 8, 2013)
19. M.P. Bacolod, J.L. Tobias, Schools, school quality and achievement growth: Evidence from the Philippines. Econ. Edu. Rev. **25**, 619–632 (2006)
20. B.K. Sovacool, S.E. Ryan, The geography of energy and education: leaders, laggards, and lessons for achieving primary and secondary school electrification. Renew. Sustain. Energy Rev. **58**, 107–123 (2016)
21. UNESCO, *Transforming Education: The Power of ICT Policies* (UNESCO, Paris, 2011)
22. M.D. Chinn, R.W. Fairlie, The determinants of the global digital divide: a crosscountry analysis of computer and internet penetration, Oxf. Econ. Pap. December 3 (2006)
23. UNESCO Teaching and Learning, *Achieving Quality for All* (UNESCO, Paris, 2014)
24. B.K. Sovacool, S. Clarke, K. Johnson, M. Crafton, J. Eidsness, D. Zoppo, The energy-enterprise-gender nexus: lessons from the multifunctional platform (MFP) in Mali. Renew. Energy **50**, 115–125 (2013)

25. A. Khellaf, F Khadri, S. Guezzane, Photovoltaic option in Algeria: current action, in *Euro-Mediterranean Workshop on Euro-Mediterranean Renewable Energies* (European Commission's INCO Program, Nicosia, Cyprus, May 18–21, 2002)
26. T. Bernard, Impact Analysis of rural electrification projects in Sub-Saharan Africa. World Bank Res. Observer **27**(1), 33–51 (2012)
27. IEA, World Energy outlook 2014, Traditional use of Biomass for cooking (2014)
28. J. Ter-Wengel, The effects of electrification and the extension of education on the retention of population in rural areas of Colombia, in *Impact of Rural Development Projects on Demographic Behaviour*, ed. by R.E. Bilsborrow, P.F. DeLargy (New York, United Nations Fund for Population Activities, New York, 1985), pp. 47–64
29. A. Gurung, Om Prakash Gurung, Sang Eun Oh, the potential of a renewable energy technology for rural electrification in Nepal: a case study from Tangting. Renew. Energy **36**, 3203–3210 (2011)

Chapter 6
Promoting Rural Electrification in Sub-Saharan Africa: Least-Cost Modelling of Decentralized Energy-Water-Food Systems: Case Study of St. Rupert Mayer, Zimbabwe

J. Winklmaier and S. Bazan Santos

Abstract The outstanding solar potential in Sub-Saharan Africa (SSA) enables significantly cheaper levelized costs of electricity for decentral solar systems compared to the commonly used diesel generators. Yet, the limited purchase power in SSA impedes rural electrification by solar systems due to their high investment costs. Decentralized Energy-Water-Food systems (EWFS) have the potential to solve this problem. Using solar-powered water pumps, rural communities can supply water for drinking and irrigation. Thereby, agriculture does not depend on rainfall solely and can be done all over the year, which leads to increasing productivity. The increased crop production reduces the community's expenses for nutrition and enables profit by sales, which in turn enables a payback of the initial investment costs of the solar system. The increased amount of biomass waste enables economically feasible small-scale biogas production. The biogas can be used for electricity production by biogas motors. These can supply private, social or small commercial loads, which enhance the local productivity even more. To identify the least-cost system design regarding the supply of electricity, water and food for the rural village of St. Rupert Mayer, Zimbabwe, the linear optimization model *urbs* was adapted. *urbs* was developed for energy system modelling, yet its sector coupling feature allows to add processes like water pumps and commodities such as biogas. The modelling results show that a holistic system including photovoltaics (PV), water pumps, enhanced agriculture and biogas production reduces the levelized costs of electricity (LCOE) from 0.45 USD/kWh by power supply from diesel generators to 0.16 USD/kWh. The modelling results shall support local governments and entrepreneurs in their decision-making.

Keywords Decentralized renewable energy solutions · Energy-Water-Food systems · Biogas · Least-cost modelling · Economic development

---

J. Winklmaier (✉) · S. Bazan Santos
Technical University of Munich, Lichtenbergstrasse 4a, 85748 Garching, Germany
e-mail: johannes.winklmaier@tum.de

M. Mpholo et al. (eds.), *Africa-EU Renewable Energy Research and Innovation Symposium 2018 (RERIS 2018)*, Springer Proceedings in Energy,
https://doi.org/10.1007/978-3-319-93438-9_6

6.1 Introduction

"Access to affordable, reliable, sustainable and modern energy for all" is one of the 17 sustainable development goals (SDG) set by the United Nations [1]. Economic growth is hindered by insufficient, unreliable and expensive availability of electricity and is felt by consumers of all sizes [2]. However, at this moment a total of 1.3 billion people worldwide have no access to electricity, 600 million of whom live in Sub-Saharan Africa (SSA) [3]. While the extraordinary potential for solar energy in SSA allows for a low levelized cost of electricity (LCOE) by means of photovoltaics (PV), which is lower than electricity generated by conventional fossil fuels, the share of solar and wind energy in primary energy consumption in SSA is currently smaller than 0.1% [3]. A major reason for this is the high interest rates typically found in SSA in combination with low purchasing power of the masses. These have an especially negative impact on investments in volatile renewable energy (RE) technologies since these have typically higher specific investment costs than conventional fossil fuel powered energy production units. This negative effect of limited purchasing power on the implementation of RE systems is further aggravated by the additional need for investment in storage and control systems. Thus, in off-grid regions the loads have to be optimized to minimize storage costs while increasing the local productivity to justify the investment in PV. Solar water pumps are a good example of such loads as they can be used flexibly and add to agricultural productivity by the pumped water. Another advantage of these decentralized Energy-Water-Food systems (EWFS) is that the resulting biomass residue can produce cost-effective biogas. This can be used to compensate the volatile PV electricity production using cheap generators.

6.2 Literature Review

The conceptualization of decentralized EWFS was driven by large-scale studies on software-based optimal planning and sizing of hybrid renewable energy systems (HRES), which propose the integration of renewable, traditional energy technologies and energy storages [4]. The complexity of these integrated energy supply systems is approached through cost-optimal modelling, simulation and optimization tools such as SOLSTOR, SOLEM and HOMER, reviewed in [5]. These tools dimension energy systems based on economic criteria as the total net present cost (TNPC) or levelized cost of energy (LCOE). The aim is to propose cost-feasible system configurations in contrast to other software tools, which optimize technical parameters such as the overall system efficiency [6]. Cost-optimal HRES are the most widely spread state-of-the-art solution for sustainable rural development through reliable and cost-effective electricity supply [5]. The economic feasibility of the proposed system designs has been positively measured through the impact on the agricultural sector. Agriculture is one main primary source of income and of local food supply [7]. By performing pre-analysis and calculations related to agricultural processes outside the optimization tools, the model outputs fit closely to the rural reality. Hence, several studies assess load requirements for agricultural irrigation apart from the domestic load [8, 9]. Another approach is the scheduling of daily irrigation according to the operation of photovoltaic-powered water pumping systems [10, 11]. Research is also effectuated on the usage of crop residues to power cost-free fuelled biomass generators [12] or the generation of biogas through anaerobic fermentation as well as from crop residues use in order to power biogas generators [13, 14]. These pre-analysis calculations carried out before the simulation address the dependency of agricultural processes (water and food systems) on energy technologies, portrait the local eco-system more competently and assess rural development goals.

However, a key desirable feature is the direct modelling and simulation of processes relevant to EWFS in cost-optimizing HRES software in order to properly dimension the integrative decentralized system. Studies approaching the EWF nexus highlight the need for integrative solutions that consider the synergies among the components of EWFS and assess the connections between the three subsystems [15, 16]. Existing tools address the interdependencies of these complex systems from project management-based frameworks [17] or techno-ecological methodologies like resource management [16]. Hence, this research aims to contribute to the current research status-quo by assessing the economic feasibility of EWFS with least-cost modelling. This is achieved with *urbs*, a linear optimization model developed at the Chair of Renewable and Sustainable Energy Systems, Technical University Munich (TUM), Germany. All previous studies on *urbs* have focused on grid-connected [18, 19] or off-grid [20] energy systems. *urbs* allows, however, the modelling of cross-sector systems as transformation processes of input and output resources and the optimization of the integrated system. The results show the optimal planning and operation for the holistic EWFS. Furthermore, the trade-offs across the energy-water-food nexus and the associated impact can be analysed. The workflow is outlined in the next section.

6.3 Methodology

This study uses the model *urbs* for the analysis of the EWF system scenarios in decentralized off-grid rural areas. *urbs* is an open-source linear optimization tool programmed in the language Python. It is an economic model and identifies the optimal system configuration that meets a predetermined resource demand with respect to the economic feasibility. This is given by the minimal variable total costs resulting from the techno-economic modelling of each process, transmission and storage technologies in the system. Furthermore, *urbs* offers features for the design of buying and selling processes, intertemporal optimization as well as demand-side

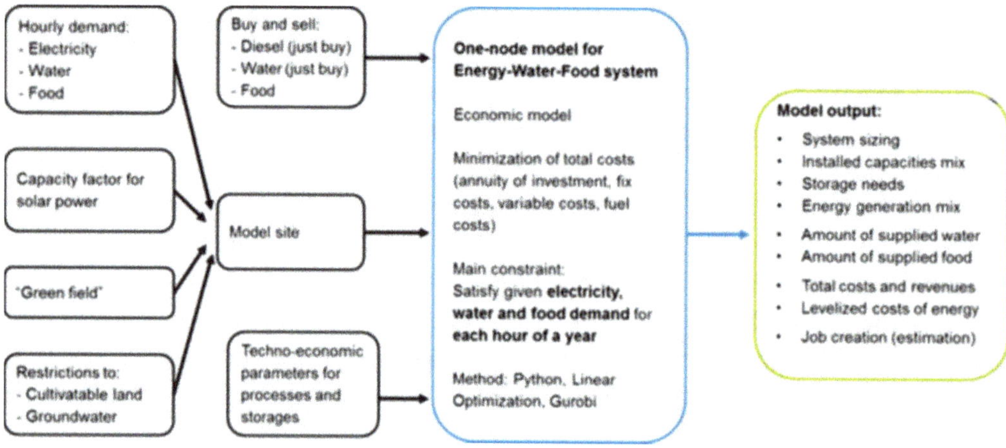

Fig. 6.1 One-node modelling of EWFS with *urbs*

management, and allows configurable time and space resolution (one-node or multiple-node system) [18]. The complete mathematical description of *urbs* can be found in [21].

Figure 6.1 shows the schematic EWFS framework with the main input and outputs. This work focuses on the off-grid one-node system, characterized by local weather conditions, demand, resource availability and site topography. There are market-based prices for purchasing and selling of selected resources and all EWF processes and storages are subject to local techno-economic parameters. The model outputs listed in Fig. 6.1 are obtained under the main restriction of satisfying the hourly EWF demands. The proposed model is tested on a Zimbabwean use case village.

St. Rupert Mayer (SRM), a village in rural Zimbabwe with 250 inhabitants, was chosen as a case study for the modelling of an EWFS was chosen. The long-term cooperation between SRM and TUM facilitated the acquisition of required input data for the model regarding demand and supply of power, water, agriculture and food. Based on the nominal power of residential applications and their temporal utilization patterns, a Monte Carlo Simulation was conducted to create an hourly electrical load profile for SRM [22]. The daily water demand is 50 l per person, the daily food demand is represented by 600 g wheat per person, which covers the required 2200 kcal per day [1]. The demand time series for power, water and food include just residential consumption, additional loads such as the power demand by water pumps are implemented as possible processes. Because of the unreliability of the grid connection, SRM is assumed to be off-grid regarding the modelling. Power supply is possible by diesel generators, biogas generators, PV and batteries. The time series for global horizontal radiation in SRM has been created with "*PVWatts*" and converted into hourly capacity factors for PV [22]. Water can be purchased by water trucks for 0.80 USD/m^3 or supplied by water pumps and stored in water tanks. Wheat can be purchased and sold on the local market for each 0.45 USD/kg or supplied by local farming. Because of the missing water grid connection and the unreliable power grid, presently farming in SRM is just possible in small scales

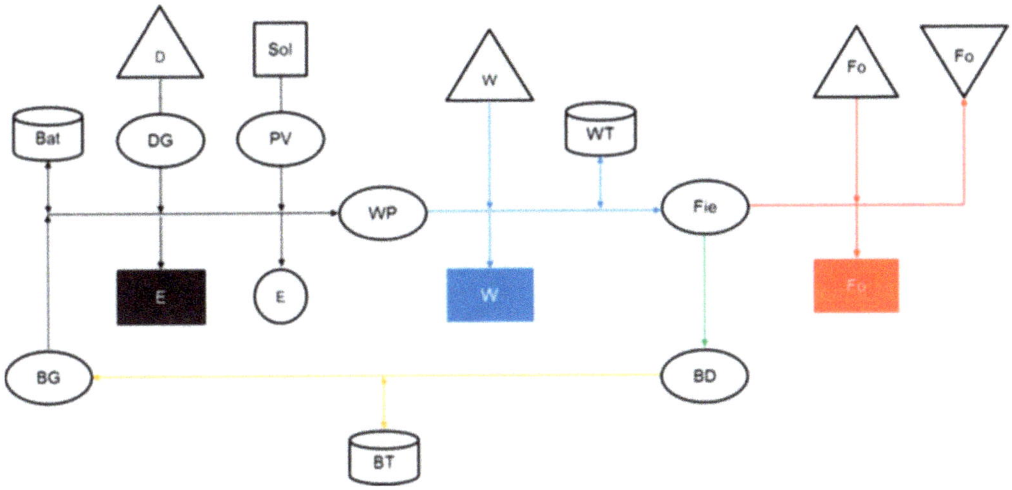

Fig. 6.2 Scheme of EWFS

during rainy season. In the model, just farming by means of groundwater pumping is taken into account. The wheat straw can be used as free fuel for a biogas digester in order to generate fuel for the biogas generator. Based on questionnaires, the variable costs of the process *Field* represent an annual income of 750 USD for farmers in SRM. The maximum capacity for farming is set to 10 ha and for groundwater to 300,000 m^3 per year, both based on values during peak productivity in the 1990s. Although SRM is modelled as a "green field" in order to give generic results, all described Energy-Water-Food processes are or have been implemented in SRM. Thus, it was possible to acquire the locally occurring CAPEX and OPEX for these processes by questionnaires in SRM [23]. General techno-economic process parameters are defined based on literature research, e.g. the weighted average cost of capital (WACC) is set to 15%. All model assumptions are listed in the appendix.

The investigated scenarios are distinguished by the amount of implemented demand-commodities, processes and storages. Scenario 1 represents a simple diesel island system, scenario 9 includes all possible processes of the modelled EWFS as shown in Fig. 6.2 (Table 6.1).

6.4 Results and Discussion

Figure 6.3 shows a result plot for the least-cost power generation with *urbs*. The black line is showing the implemented power demand, the additional demand due to the water pump is visualised as a negative power producer. The power producers *Diesel Generator* and *Photovoltaics* generate sufficient electricity to cover both types of demand including storage losses. The excessive power being produced by PV is causing dump load, "*Shunt(Elec.)*", which is shown as a negative producer as well.

Table 6.1 Modelled scenarios

#	Title	Modelled demands, processes, and storages etc.
S1	Diesel	Electricity (E) *demand* + Diesel (D) *buy* + Diesel Generator (DG)
S2	+PV	S1 + Photovoltaic (PV) + Electricity (E) *dump load*
S3	+Battery	S2 + Battery (Bat)
S4	+Water Demand	S3 + Water (W) *demand* + *buy*
S5	+Water Pump	S4 + Water Pump (WP) + Water Tank (WT)
S6	+Food Demand	S5 + Food (Fo) *demand* + *buy*
S7	+Fields	S6 + Field (Fie)
S8	+Food Selling	S7 + Food (Fo) *sell*
S9	+Biogas	S8 + Biogas Digester (BD) + Biogas Tank (BT) + Biogas Generator (BG)

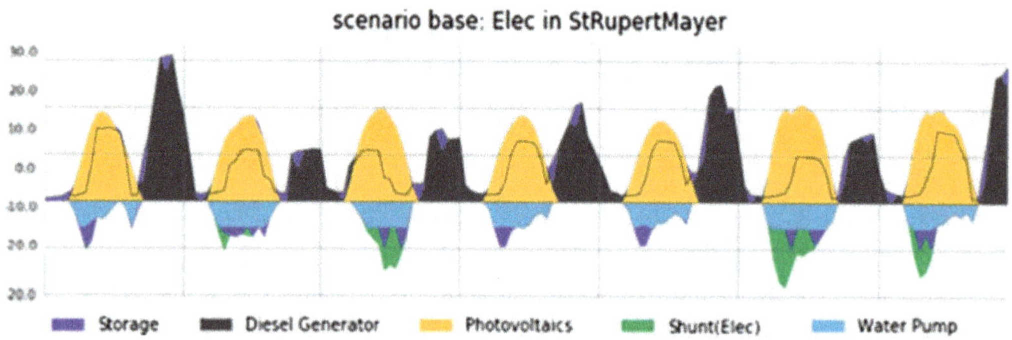

Fig. 6.3 Power generation time series for one week in kW—Scenario "+*Fields*"

Figure 6.4 shows the least-cost power generation mix for each scenario. As soon as PV is allowed to be installed from scenario "+*PV*" on, it generates a relevant share of power, yet the Diesel generator continues to supply power during night hours. For all scenarios including power generation with PV, the least-cost generation includes overproduction by PV and thus dump load, even if batteries are allowed, which can be observed in Fig. 6.3. The use of water pumps leads to increasing PV capacity and decreasing battery capacity or dump load because the water pump is working as a flexible load. This effect is minor for the supply of private water demand in scenario "+*Water pump*", yet increases significantly in case of increased water supply for agriculture in the scenarios "+*Fields*" and "+*Food selling*". In this scenario, the maximum available land of 10 ha is completely used. In scenario "+*Biogas*" there is no more power generation from PV and just a slight generation from diesel generators. The selling-oriented agriculture results in 342 t of straw, which can be used as free fuel for the biogas digester and its energy content of 530 MWh exceeds the total power demand of the EWFS by far.

As shown in Fig. 6.5, the LCOE decrease from 0.45 to 0.38 USD/kWh if power can be generated by diesel generators and PV. The introduction of batteries and

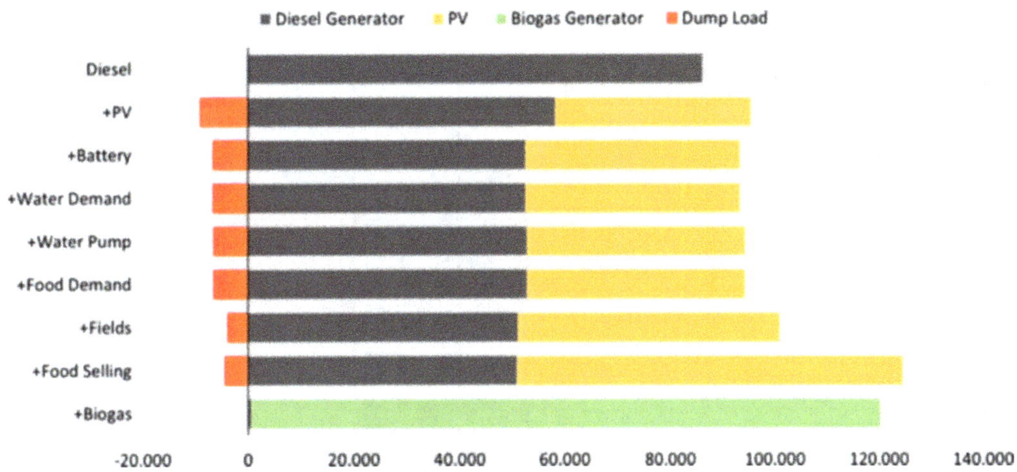

Fig. 6.4 Annual power generation in kWh for each scenario

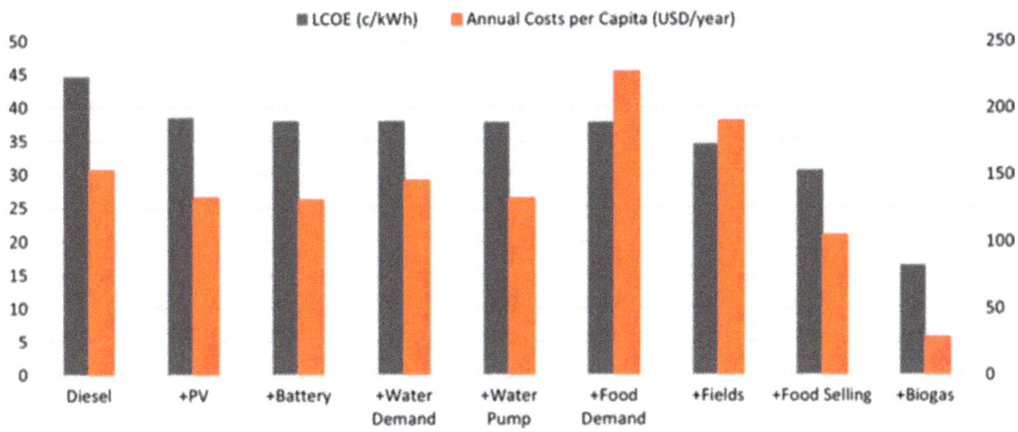

Fig. 6.5 LCOE and annual costs per capita for each scenario

water pumps for private water demand reduces the LCOE just minimally. Yet, the increased water pump capacities of the scenarios "+*Fields*" and "+*Food Selling*" reduce the LCOE significantly to 0.34 and 0.31 USD/kWh, respectively. This is due their cost-efficient utilization during the day, as shown in Fig. 6.3. The introduction of biogas leads to a further reduction of the LCOE to 0.16 USD/kWh, because of the moderate CAPEX of both processes, low storage costs for the biogas tank and the sufficient availability of straw as free input of the digester. The described reductions of LCOE lead to reductions of annual costs per capita, as it can be seen in Fig. 6.5. However, the annual costs rise in the scenario "+Water demand" and especially in the scenario "+Food demand" because of the additional costs for water and food, which have to be purchased as shown in Fig. 6.6. The scenarios "+*Water Pump*" and "+*Fields*" show significantly lower annual costs than these because water and food are now self-supplied by water pumps and fields on lower costs. The revenue by selling wheat in the last two scenarios leads to a drastic cost reduction.

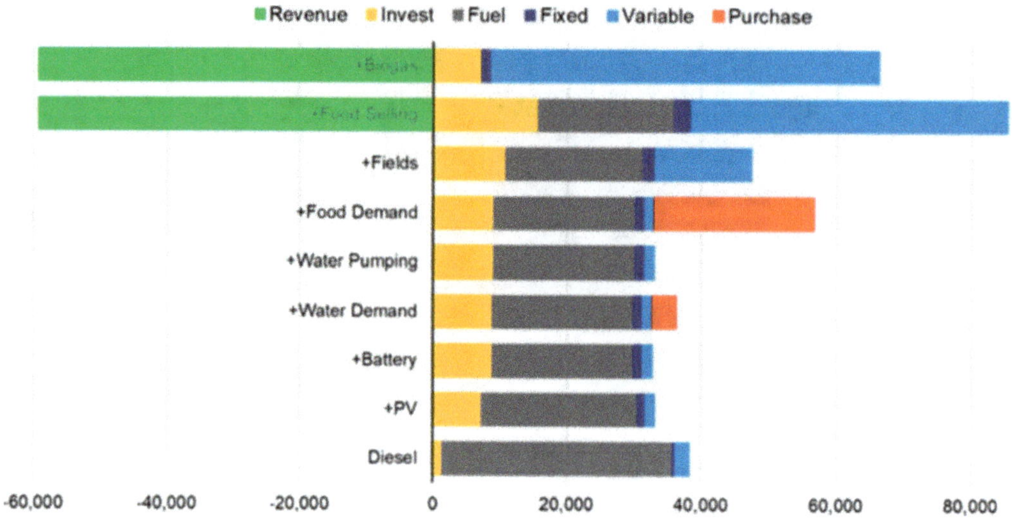

Fig. 6.6 Annual system costs in USD per year for each scenario

29 USD per year and person in scenario "+*Biogas*" is remarkable, because these expenses cover the consumption of power, water and food per capita. The identical demand causes annual costs of 227 USD per person in scenario "+*Food demand*". The expenses on fuel and purchase of food dominate this scenario's costs, but there is no revenue from sales. The system costs in the last scenario are actually higher than the ones of scenario "+Food demand", yet almost offset by the revenue of the wheat sales. Furthermore, its system costs are not caused by fuel and purchase but are dominated by variable costs. These are caused by the high maintenance costs of the biogas digestion and in particular the salary for farmers, which also occur in scenario "+*Fields*" and "+*Food selling*". Based on the resulting variable costs, this scenario leads to the creation of 60 jobs in agriculture. The job creation could be even higher if more cultivatable land would be available. Figure 6.6 shows also, that in no scenario the CAPEX ("*invest*") are dominating, mostly due to the purchasing costs for diesel. In the last scenario, no investment-intensive PV is installed anymore.

6.5 Conclusion and Recommendations

The results show the techno-economic benefits of decentralized EWFS. The use of solar-powered water pumps enables sufficient water supply at low costs for enhanced agriculture. The increased agricultural productivity leads to revenues by crop sales, job creation in farming, and big quantities of crop waste like straw, which can be converted on low costs to biogas and thus power the whole system. Both the biogas generation and utilization as well as the revenues by crop sales reduce the total system costs for the supply of power, water and food to less than 11% of the costs occurring in case of power generation with diesel and water and food purchase from the market. Additionally, the job creation in agriculture creates local purchase power, which enables to pay off the investment costs of the system.

Further model improvement will address: Implementation of rainfall time series, modelling groundwater as storage, enhanced soil analysis, fertilizer demand, fertilizer supply by commercial ones or residues from biogas digester, life-cycle-analysis for soil nutrients, additional field crops, variation of crop prices, livestock farming, food processing, other electrical loads from commercial and social sector, heat demand for cooking, heat supply by firewood and biogas, labour as commodity.

Additionally, the obtained results have to be implemented to identify obstacles, which are not covered by the model yet, such as required technical expertise. Based on these findings, governments, entrepreneurs and other local stakeholders shall be able to promote electrification and economic development of rural regions in SSA by means of EWFS.

Appendix 1: Techno-Economic Input Parameter

See Tables 6.2, 6.3, 6.4, 6.5, 6.6, 6.7, 6.8, 6.9, and 6.10.

Table 6.2 Techno-economic parameters for diesel generator [20, 24, 25]

	Unit	Value
Investment costs	US$/KW	240
Fixed O&M costs	US$/KW/year	5% of investment costs
Variable O&M costs	US$/kWh	0.024
Diesel price	US$/kWh	0.1
Load efficiency	%	25
Depreciation time	year	20

Table 6.3 Techno-economic parameters for solar photovoltaics [20, 26]

	Unit	Value
Investment costs	US$/kW	1600
Fixed O&M costs	US$/kW/year	2% of investment costs
Modular efficiency	%	16.7
Depreciation time	year	20

Table 6.4 Techno-economic parameters for water pumps [27, 28]

	Unit	Value
Investment costs	US$/kW	900
Fixed O&M costs	US$/kW/year	5% of investment cost
Total dynamic head	m	70
Pump efficiency	%	80
Depreciation time	year	15

Table 6.5 Techno-economic parameters for wheat fields [29–31, 39, 40]

	Unit	Value
Max capacity of fields	ha	10
Variable O&M costs	US$/ton	247.73
Average wheat yield	ton/ha	6
Wheat growth time	day	120
Max cycles per year	–	3
Labor requirements	farmer/ha	6
Wheat ratio to wheat plant	%	35
Wheat waste ratio to wheat plant	%	65
Wheat plant water requirements	m^3/ton	263.58
Depreciation time	year	20

Table 6.6 Techno-economic parameters for biogas digester [32–34, 36, 41, 42]

	Unit	Value
Investment costs	US$/kW	500
Fixed O&M costs	US$/kW/year	2% of investment costs
Variable O&M costs	US$/kWh	0.024
Calorific energy in 1 m^3 biogas	kWh	6.2
Biogas yield from wheat	m^3/ton	450
Biogas yield from wheat waste	m^3/ton	250
Depreciation time	year	20

Table 6.7 Techno-economic parameters for lithium battery [20, 35]

	Unit	Value
Investment cost (Energy)	US$/kW	250
Investment cost (Power)	US$/kWh$_{el}$	350
Fixed cost (Power)	US$/kW/year	2% of investment costs (Power)
Fixed cost (Energy)	US$/kWh$_{el}$/year	10
Variable cost (Power)	US$/kWh$_{el}$	0.02
Input efficiency	%	95
Output efficiency	%	95
Initial state of charge (SOC)	%	50
Depreciation time	year	10

Table 6.8 Techno-economic parameters for biogas generator [36–38, 43]

	Unit	Value
Investment costs	US$/kW	400
Fixed O&M costs	US$/kW/year	5% of investment costs
Variable O&M costs	US$/kWh$_{el}$	0.02
Efficiency	%	30
Depreciation time	year	20

Table 6.9 Techno-economic parameters for water tank

	Unit	Value
Investment costs	US$/m^3	30
Depreciation time	year	20

Table 6.10 Techno-economic parameters for biogas storage

	Unit	Value
Investment costs	US$/kWh	16.13
Depreciation time	year	20

Appendix 2: Time Series for Least-Cost Power Generation and Storage

See Figs. 6.7, 6.8, 6.9, 6.10, 6.11, 6.12, 6.13, 6.14, and 6.15.

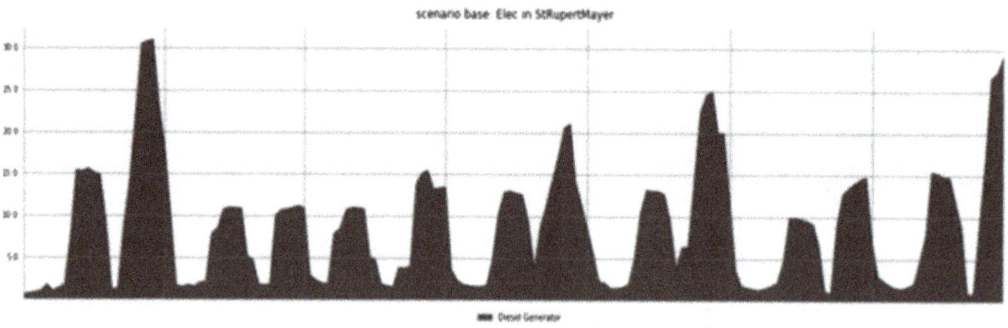

Fig. 6.7 Time series for least-cost power generation for one week in kW—Scenario "Diesel"

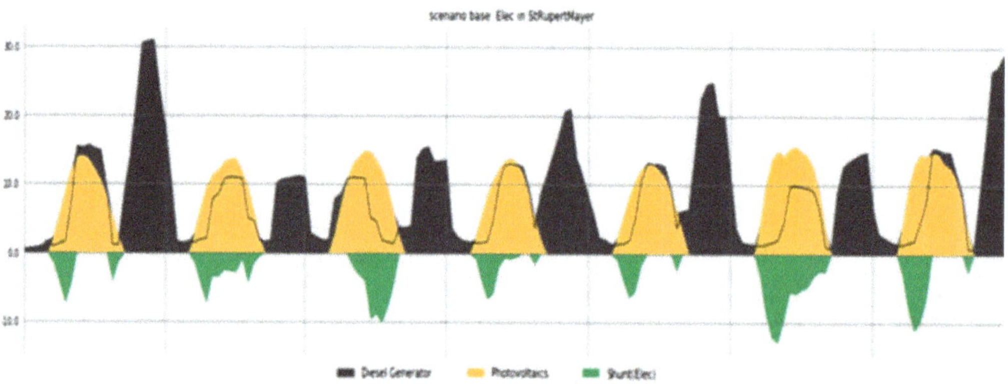

Fig. 6.8 Time series for least-cost power generation for one week in kW—Scenario "+ PV"

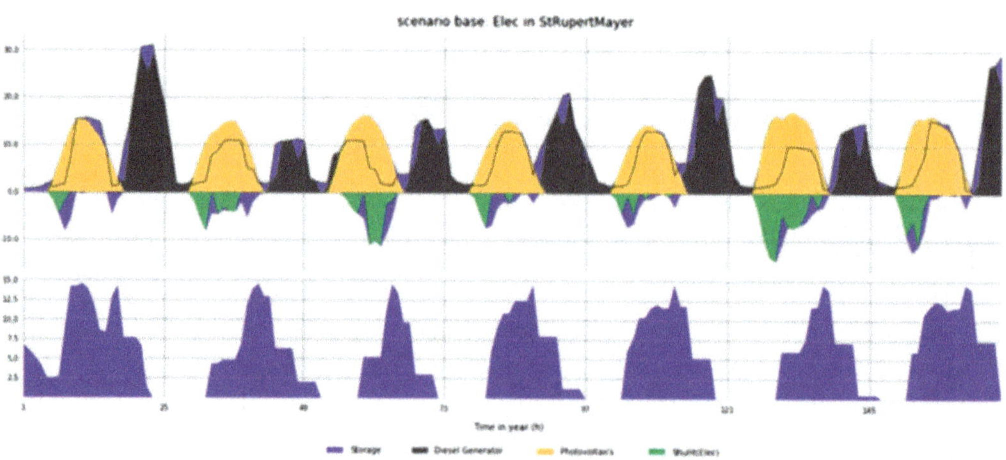

Fig. 6.9 Time series for least-cost power generation and storage for one week in kW—Scenario "+ Battery"

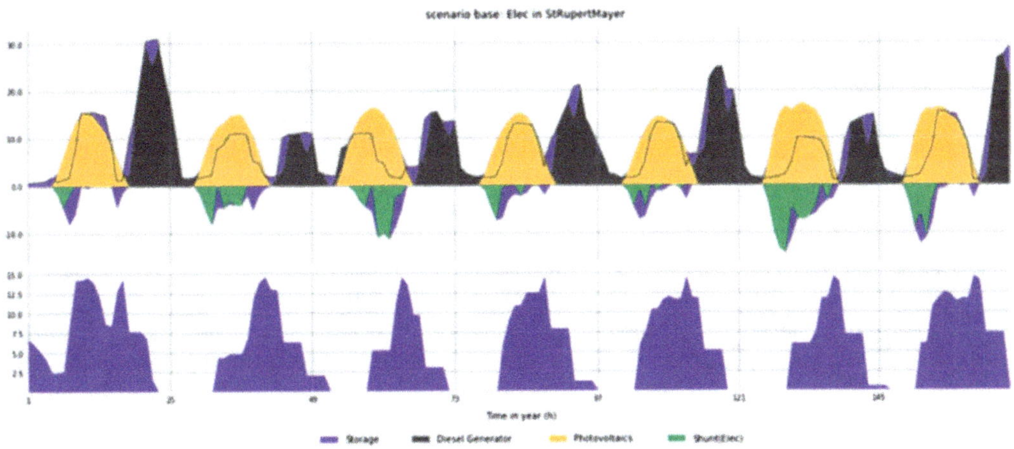

Fig. 6.10 Time series for least-cost power generation and storage for one week in kW—Scenario "+ Water Demand"

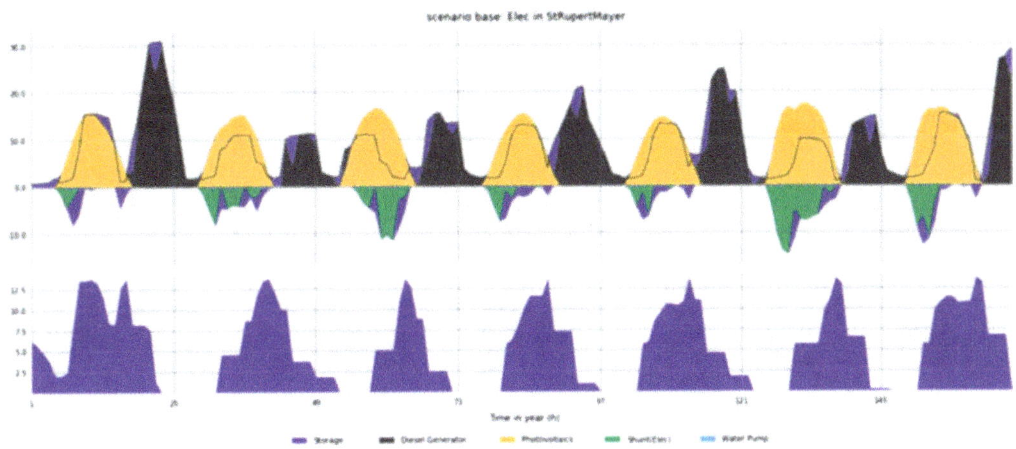

Fig. 6.11 Time series for least-cost power generation and storage for one week in kW—Scenario "+ Water Pump"

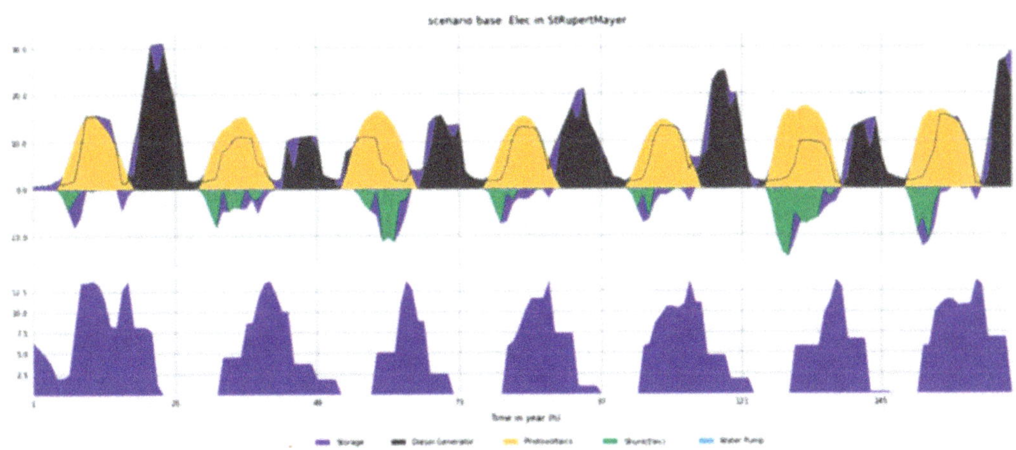

Fig. 6.12 Time series for least-cost power generation and storage for one week in kW—Scenario "+ Food Demand"

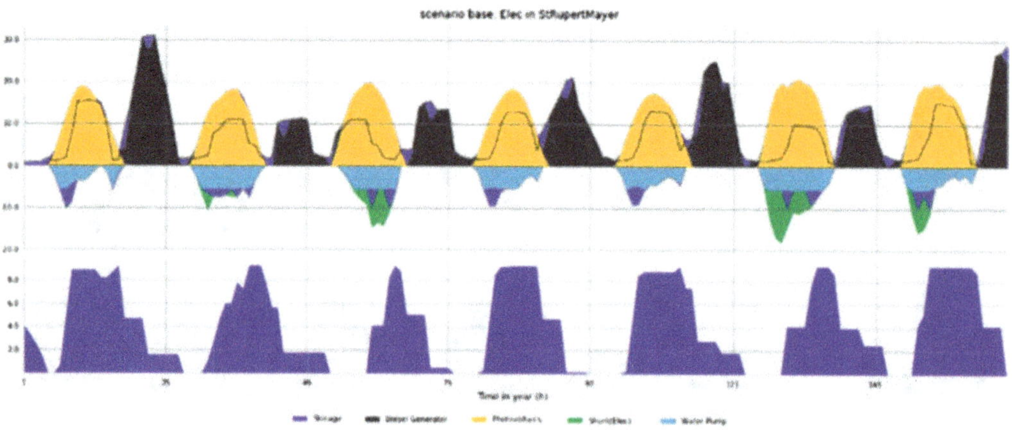

Fig. 6.13 Time series for least-cost power generation and storage for one week in kW—Scenario "+ Fields"

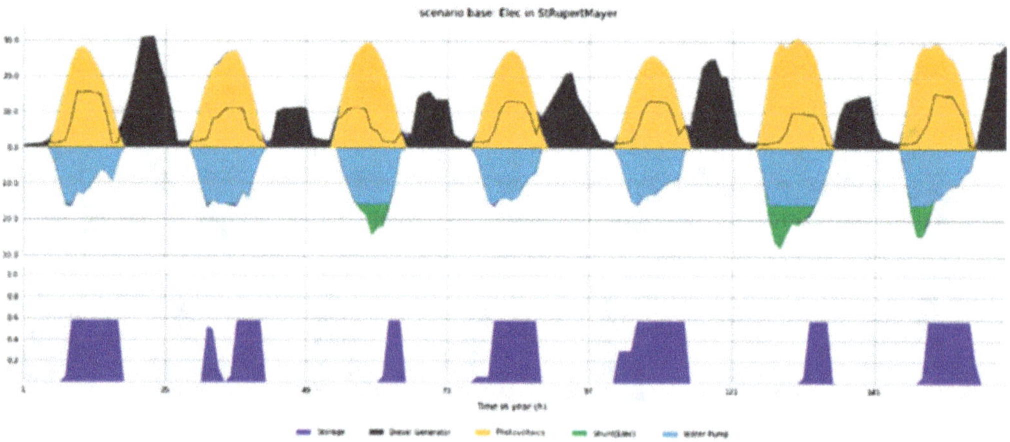

Fig. 6.14 Time series for least-cost power generation and storage for one week in kW—Scenario "+ Food selling"

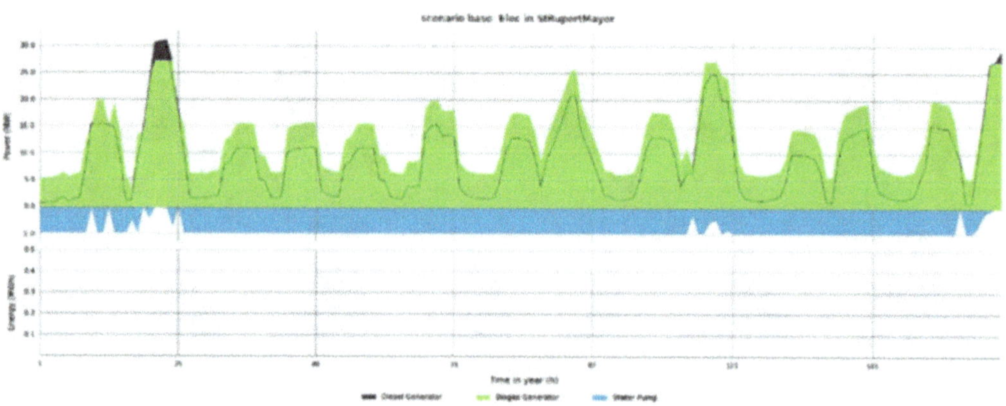

Fig. 6.15 Time series for least-cost power generation and storage for one week in kW—Scenario "+ Biogas"

Appendix 3: Output Parameter

See Tables 6.11, 6.12, 6.13, 6.14, 6.15, 6.16, and 6.17.

Table 6.11 Installed process capacities for each scenario

#	DG [kW]	PV [kW]	BG [kW]	BD [kW]	WP [kW]	Fie [ha]
S1	35.87	0	0	0	0	0
S2	35.87	23.16	0	0	0	0
S3	29.55	25.43	0	0	0	0
S4	29.55	25.43	0	0	0	0
S5	29.84	25.79	0	0	0.40	0
S6	29.84	25.79	0	0	0.40	0
S7	31.55	31.02	0	0	5.52	2.89
S8	35.32	46.03	0	0	15.68	10
S9	8.7	0	27.17	47.38	4.78	10

Table 6.12 Installed storage capacities for each scenario

#	Bat [kWh]	WT [m^3]	BT [kWh]
S1	0	0	0
S2	0	0	0
S3	14.57	0	0
S4	14.57	0	0
S5	13.68	10.48	0
S6	13.68	10.48	0
S7	9.25	9.36	0
S8	0.58	7.28	0
S9	0	0.52	222.29

Table 6.13 Energy, water and food quantities for each scenario

#	Energy [kWh]	Water [m^3]	Wheat [ton]	Straw [ton]
S1	86,104.07	0.00	0.00	0
S2	86,104.07	0.00	0.00	0
S3	86,104.07	0.00	0.00	0
S4	86,104.07	4555.20	0.00	0
S5	87,188.65	4555.20	0.00	0
S6	87,188.65	4555.20	52.56	0
S7	96,613.15	44,138.14	52.56	97.76
S8	12,0174.43	143,095.48	183.96	342.17
S9	120,174.43	143,095.48	183.96	342.17

Table 6.14 Bought and sold commodities for each scenario

#	Diesel buy [kWh]	Water buy [m^3]	Wheat buy [ton]	Wheat sell [ton]
S1	344,416.30	0	0	0
S2	232,345.65	0	0	0
S3	209,640.59	0	0	0
S4	209,640.59	4555.20	0	0
S5	211,093.85	0	0	0
S6	211,093.85	0	52.56	0
S7	204,405.85	0	0	0
S8	203,194.48	0	0	131.40
S9	2239.50	0	0	131.40

Table 6.15 Power generation mix for each scenario

#	DG [kWh]	PV [kWh]	BG [kWh]
S1	86,104.07	0.00	0
S2	58,086.41	37,130.61	0
S3	52,410.15	40,764.24	0
S4	52,410.15	40,764.24	0
S5	52,773.46	41,345.52	0
S6	52,773.46	41,345.52	0
S7	51,101.46	49,736.43	0
S8	50,798.62	73,794.97	0
S9	559.88	0.00	119,614.55

Table 6.16 System costs in USD per year for each scenario

#	Revenue	Invest	Fuel	Fixed	Variable	Purchase
S1	0	1375.39	34,441.63	430.45	2066.50	0
S2	0	7295.19	23,234.56	1171.53	1394.074	0
S3	0	8963.29	20,964.06	1345.55	1437.085	0
S4	0	8963.29	20,964.06	1345.55	1437.085	3644.16
S5	0	9102.14	21,109.39	1368.23	1431.906	0
S6	0	9102.14	21,109.39	1368.23	1431.906	23,652.00
S7	0	10,894.64	20,440.59	1733.96	14,374.95	0
S8	−59,130.00	15,635.90	20,319.45	2610.87	46,800.69	0
S9	−59,130.00	7164.73	223.95	1336.48	57,547.30	0

Table 6.17 LCOE and annual costs per capita for each scenario

#	LCOE [c/kWh]	Annual costs per capita [US$/year]
S1	44.50	153.30
S2	38.40	132.40
S3	37.90	130.80
S4	37.90	145.40
S5	37.70	132.00
S6	37.70	226.65
S7	34.40	189.78
S8	30.50	104.95
S9	16.40	28.57

References

1. United Nations, Sustainable development goals [Online]. http://www.un.org/sustainabledevelopment/sustainable-development-goals/
2. CDC, Development impact evaluation: what are the links between power, economic growth (2016)
3. International Energy Agency, Africa energy outlook, 2015
4. B. Binayak, L. Kyung-Tae, G. Lee, Y. Cho, S. Ahn: Optimization of hybrid renewable energy power systems: a review. Int. J. Precis. Eng. Manuf. Green Technol. (2015)
5. S. Yashwant, S. Gupta, A. Bohre, Review of hybrid renewable energy systems with comparative analysis of off-grid hybrid system. Renew. Sustain. Energy Rev. (2018)
6. S. Sunanda, S. Chandel, Review of software tools for hybrid renewable energy systems. Renew. Sustain. Energy Rev. (2014)
7. N. Varshney, M. Sharma, D. Khatod, Sizing of hybrid energy system using HOMER. Int. J. Emerg. Technol. Adv. Eng. (2013)
8. M. Shahzad, A. Zahid, T. Rashid, M. Rehan, M. Ali, M. Ahmad, Techno-economic feasibility analysis of a solar-biomass off grid system for the electrification of remote rural areas in Pakistan using HOMER software. Renew. Energy (2017)
9. R. Sen, S. Bhattacharyya, Off-grid electricity generation with renewable energy technologies in India: an application of HOMER. Renew. Energy (2014)
10. A. Al-Waeli, A. Al-Kabi, A. Al-Mamari, H. Kazem, M. Chaichan, Evaluation of the economic and environmental aspects of using photovoltaic water pumping system. In: 9th International Conference on Robotic, Vision, Signal Processing and Power Applications: Empowering Research and Innovation (2016)
11. M. Chaichan, A. Kazem, M. El-Din, A. Al-Kabi, A. Al-Mamari, H. Kazem, Optimum design and evaluation of solar water pumping system for rural areas. Int. J. Renew. Energy Res. (IJRER) (2017)
12. Priyanka, S. Bath, M. Rizwan, Design and optimization of RES based standalone hybrid system for remote applications. In: 2017 IEEE Power & Energy Society Innovative Smart Grid Technologies Conference (ISGT) (2017)
13. P. Ananda, S. Bathb, M. Rizwanc, Design of solar-biomass-biogas based hybrid system for rural electrification with environmental benefits. Int. J. Recent Innov. Trends Comput. Commun. (2017)
14. P. Mathema, Optimization of integrated renewable energy system—micro grid (IRES-MG). Tribhuvan University (2008)

15. UNECE. Water-food-energy-ecosystem nexus [Online]. http://www.unece.org/env/water/nexus.html
16. E. Martinez-Hernandez, M. Leach, A. Yang, Understanding water-energy-food and ecosystem interactions using the nexus simulation tool NexSym. Appl. Energy (2017)
17. H. Sieverding, D. Clay, E. Khan, J. Sivaguru, M. Pattabiraman, R. Koodali, M. Ndiva-Mongoh, J. Stone, A sustainable rural food–energy–water nexus framework for the northern great plains. Agric. Environ. Lett. (2016)
18. M. Askri, Challenges of energy system modelling for developing countries. Technical University of Munich, Institute for Renewable and Sustainable Energy Systems (2017)
19. M. Huber, A. Roger, T. Hamacher, Optimizing long-term investments for a sustainable development of the ASEAN power system. Energy (2015)
20. A. Okunlola, Assessment of decentralized energy systems in Sub-Saharan Africa: Energy system modelling and job creation analysis. Technical University of Munich, Institute for Renewable and Sustainable Energy Systems (2017)
21. urbs: a linear optimisation model for distributed energy systems [Online]. http://urbs.readthedocs.io/en/latest/
22. M. Mayr: Feasibility analysis of power supply by small-scale wind turbines in urban, semi-urban and rural districts of Zimbabwe. Technical University of Munich, Institute for Renewable and Sustainable Energy Systems (2017)
23. C. Schulze: The Nexus of Water, Food and Energy in emerging countries: Data acquisition and renewable resource potential assessment by GIS and ground measurements in Zimbabwe. Technical University of Munich, Institute for Renewable and Sustainable Energy Systems (2017)
24. GlobalPetrolPrices.com [Online]. http://www.globalpetrolprices.com/Zimbabwe/diesel_prices/
25. U.S. Energy Information Administration [Online]. https://www.eia.gov/energyexplained/index.cfm?page=about_energy_conversion_calculator
26. Cool Solar [Online]. https://www.cool-solar-africa.com/
27. Grundfos [Online]. http://magazines.grundfos.com/Grundfos/SU/UK/Groundwater/
28. A. Al-Waeli, K. Moanis, H. Kazem, M. Chaichan, Optimum design and evaluation of solar water pumping system for rural areas. Int. J. Renew. Energy Res. (2017)
29. Harvesting Crop Residues [Online]. http://extensionpublications.unl.edu/assets/pdf/g1846.pdf
30. Zimmatic [Online] Increasing Wheat Yields: Through Efficient Irrigation Solutions. http://www.zimmatic.com/stuff/contentmgr/files/0/9ee056014d476a4cfbf86883b0604ec1/pdf/z_bro_wheat__revised_9.2013_.pdf
31. CropWat [Online]. http://www.fao.org/land-water/databases-and-software/cropwat/en/
32. R. Kigozi, O. Aboyade, E. Muzenda, Technology selection of biogas digesters for OFMSW via multi-criteria decision analysis [Online] in The 2014 International Conference of Manufacturing Engineering and Engineering Management
33. Biogas in India [Online]. http://large.stanford.edu/courses/2010/ph240/pydipati2/
34. Cropgen [Online]. http://www.cropgen.soton.ac.uk/deliverables.htm
35. H. Hesse, Elektrische Energiespeicher für stationäre Anwendungsfälle (2017)
36. P. Mukumba, G. Makaka, S. Mamphweli, S. Misi, A possible design and justification for a biogas plant at Nyazura Adventist High School, Rusape, Zimbabwe. J. Energy South. Afr. (2013)
37. S. Mandal, H. Yasmin, M. Sarker, M. Beg, Prospect of solar-PV/biogas/diesel generator hybrid energy system of an off-grid area in Bangladesh [Online], in AIP Conference Proceedings 1919 (2017)
38. R. Rownak, K. Ahmed, M. Shajibul-Al-Rajib, Solar-Biomass-CAES hybrid system: proposal for rural commercial electrification in Bangladesh. Int. J. Innov. Appl. Stud. ISSN 19 (2017)
39. Management of irrigated wheat [Online]. http://www.fao.org/docrep/006/y4011e/y4011e0r.htm
40. Seed Co wheat varieties on the market [Online]. http://www.seedcogroup.com/zw/media/blog/seed-co-wheat-varieties-market

41. S. Ahmad, K. Mahmood, A. Muhammad, Designing and strategic cost estimation of biogas plant: an alternative for current energy crisis in Pakistan. Int. J. Renew. Energy Environ. Eng. (2015)
42. B. Amigun, H. von Blottnitz, Capacity-cost and location-cost analyses for biogas plants in Africa. Resour. Conserv. Recycl. (2010)
43. S. Sigarchian, R. Paleta, A. Malmquist, A. Pina, Feasibility study of using a biogas engine as backup in a decentralized hybrid (PV/wind/battery) power generation system—Case study Kenya. Energy (2015)

Chapter 7
Remote Sensing Techniques for Village Identification: Improved Electrification Planning for Zambia

Catherina Cader, Alin Radu, Paul Bertheau and Philipp Blechinger

Abstract Access to energy remains a challenge in many regions of Africa. In Zambia, only approximately 28% of the total population of 17 million has access to electricity, with even lower access rates of 5% in rural areas. One of the first challenges in providing these regions with reliable electricity is identifying the location of small settlements that still lack access to electricity. Systematic electrification planning requires in detail information about the current extent of electrification and the spatial location and distribution of villages and households without access. If this is available, transparent planning mechanisms can assess different electrification options such as stand-alone systems, mini-grids or grid extension. This paper aims at detecting a sample of Zambian villages without electricity through remote sensing techniques. These techniques involve applying various machine learning algorithms to classify medium resolution Sentinel 2 multispectral imagery. As results show, it is possible to identify the location and spatial extension of rural settlements in the research area. However, for an accurate assessment of the population in the respective areas, more information, such as population distribution is needed. These results will support official bodies such as the Rural Electrification Authority (REA) as well as private project developers with an entrepreneurial interest in the region. Thereby, this knowledge enables improved legal and regulatory decisions as well as increased private sector participation.

Keywords Remote sensing · Village detection · Spatial planning
Energy access · Zambia

C. Cader (✉) · A. Radu · P. Bertheau · P. Blechinger
Reiner Lemoine Institut, Rudower Chaussee 12, 12489 Berlin, Germany
e-mail: catherina.cader@rl-institut.de

A. Radu
e-mail: alinradu86@gmail.com

P. Bertheau
e-mail: paul.bertheau@rl-institut.de

P. Blechinger
e-mail: philipp.blechinger@rl-institut.de

M. Mpholo et al. (eds.), *Africa-EU Renewable Energy Research and Innovation Symposium 2018 (RERIS 2018)*, Springer Proceedings in Energy,
https://doi.org/10.1007/978-3-319-93438-9_7

7.1 Introduction

Although the situation in Sub Saharan Africa has strongly improved in recent years, lack of access to electricity is still a problem. In 2016, only 43% of the population in the region had access to electricity. In Zambia, the situation is no exception, with two thirds of the 17 million people lacking electricity [1]. Systematic electrification planning requires in detail information about the current extent of electrification and the spatial location and distribution of villages and households without access to reliable electricity. If this is available, transparent planning mechanisms can assess different electrification options such as stand-alone systems, mini-grids or grid extension [2, 3].

Remote sensing involves applying machine learning algorithms to classify medium resolution Sentinel 2 multispectral imagery [4]. This method provides quick and accurate results on large areas with minimum costs, since many satellite imagery providers offer their products for free, as well as a significant amount of open source satellite image processing software exists. As a final product we aim to identify the location and spatial extent of rural settlements in the research area. Combining this with traceable night light emissions on satellite imagery and infrastructure data will reveal the level of energy access of each settlement. These results may support official bodies such as the Rural Electrification Authority (REA) as well as private project developers with an entrepreneurial interest in the region. Thereby, this knowledge could enable improved legal and regulatory decisions as well as increased private sector participation.

7.2 Remote Sensing as a Solution

For this analysis, a combination of remote sensing and GIS tools is used (Fig. 7.1) to identify and then quantify the spatial extent of inhabited areas in wards in the Northern Province of Zambia. For this purpose we used Level 2A [5] multispectral images of the Sentinel-2 mission. Due to their higher resolution of 10 m, only four of the 12 bands provided were considered: Blue (with a wavelength of 490 nm), Green (560 nm), Red (665 nm) and Near Infrared (835 nm). The other geographic and demographic data (ward shapefiles, population count and density per ward) were collected from secondary sources [6].

The data was analyzed using the Semi-Automatic Classification Plugin (SCP) for QGIS [7] and various other QGIS tools (clip raster, raster to vector). The processing was carried out using the Minimum Distance Classification (MDC) algorithm.

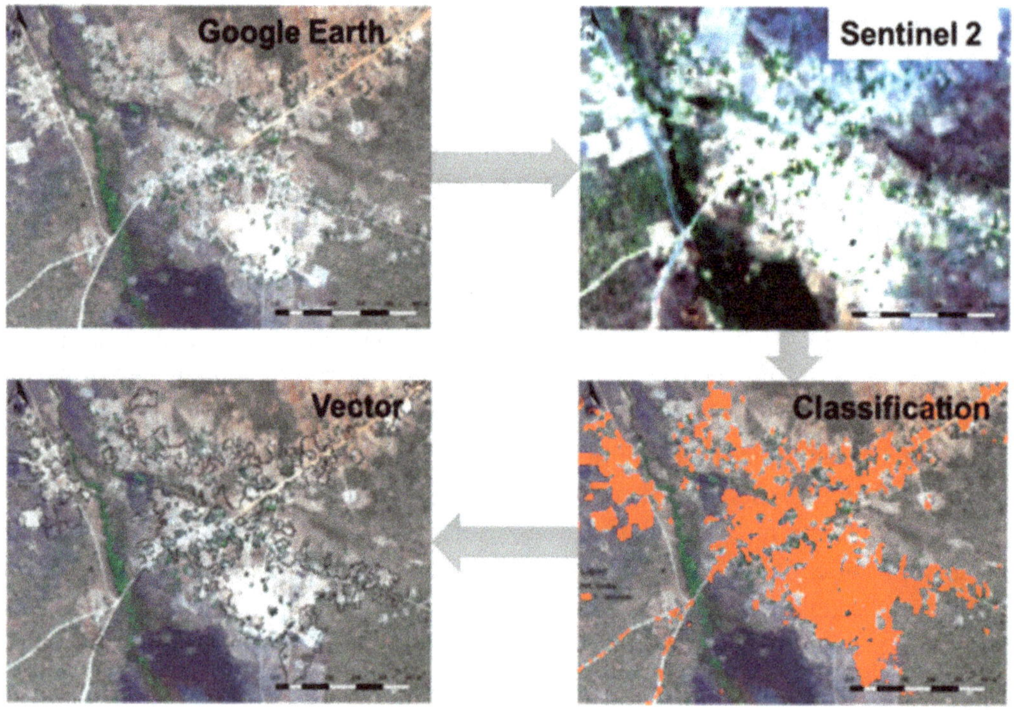

Fig. 7.1 Example of satellite image analysis (based on Copernicus Sentinel data 2017) overlayed on Google Satellite (© 2018 TerraMetrics)

7.3 Methods

The cloud free images were downloaded from the European Space Agency SciHub, for the months of July, September and October 2017.

In order to improve accuracy of classification results, some preprocessing steps need to be taken [8]. The pixel information was converted to Top of Atmosphere Reflectance and Dark Object Subtraction was applied to each Sentinel-2 scene, using SCP. As a second preprocessing step, scenes were clipped at ward level, in order to minimize areas of interest for faster processing.

Processing of each clipped ward consists of two steps: training area input and the application of a classification algorithm. The training areas are manually defined by the user, based on the visual inspection of the land cover. Classes are then defined accordingly: water, swamp, bare land, built-up, beach, vegetation and the Minimum Distance Classifier is applied. It calculates the Euclidean distance between the spectral signatures of the training input areas and the image pixels, assigning the class of the closest spectral signature [7].

After the classification of all wards, post-processing steps are required. The classification raster files obtained should be transformed to vectors, in order to easily eliminate irrelevant classes and small polygons.

7.4 Results

All the resulted vector files (each for every ward) were then merged, in order to get an overview of the whole area (Fig. 7.2).

The filtering process eliminated all polygons that had a surface area of less than 1000 m^2, which is the equivalent of 10 pixels and thereby eliminates small irrelevant regions.

In areas covered mainly by vegetation, the overall accuracy was calculated at over 90% (Fig. 7.3). However, it has been observed that in areas which contain swamp or sandy beaches, the built-up class is overestimated due to the similarity of spectral signatures, resulting in a drop in accuracy to a minimum of approximately 50% (Fig. 7.4). Accuracy can be improved by the use of high (5 m) resolution satellite imagery, such as RapidEye [9], which provides their data at no cost for academic or non-profit purposes.

A number of 14 wards were classified, with a total surface area of approximately 10,000 km^2 and, according to citypopulation.de, a total population of almost 90,000 inhabitants [6]. From the total surface area, 0.05% was attributed to the "built-up" class. The resulting file which indicates the location and spatial extent of villages could be overlaid on grid data, in order to get a better idea of the village distance to the grid.

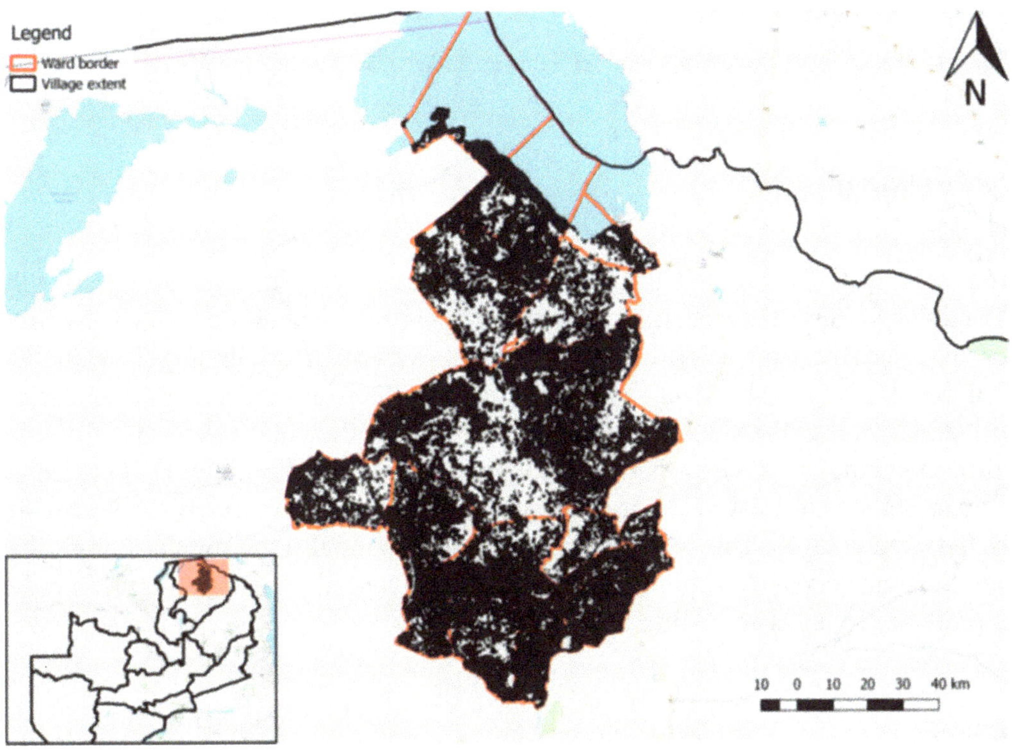

Fig. 7.2 Overview of the classified wards [overlaid on Open Street Maps © OpenStreetMap contributors under Open Data Commons Open Database License (ODbL)]

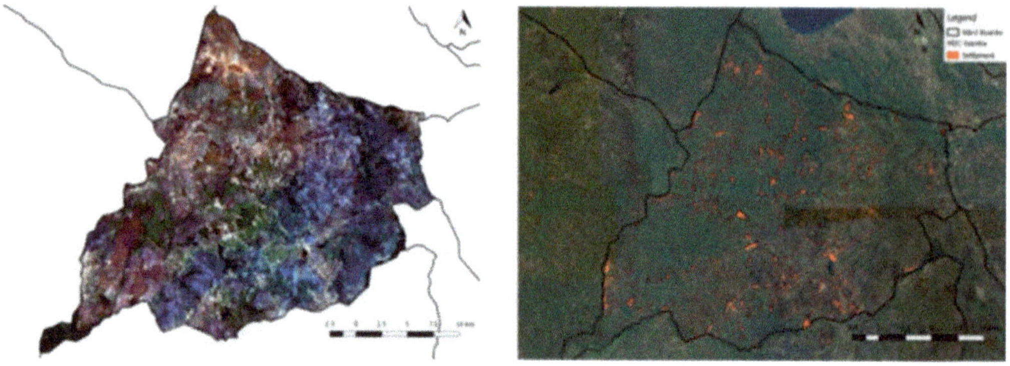

Fig. 7.3 Vyamba ward, an example of high accuracy village detection analysis (based on Copernicus Sentinel data 2017), overlaid on Google Satellite (© 2018 TerraMetrics)

Fig. 7.4 Malalo ward, overestimation of built-up areas in a swamp environment (based on Copernicus Sentinel data 2017)

7.5 Conclusion

Remote sensing can be used as a tool for the detection of built-up areas, specifically villages, regarding their position and extent. This information could then be used for the electrification prioritization of villages, through combining it with, for example, grid data. This data can then be used in decisions such as grid extension (if the village is close to the grid) versus mini-grid (if the village is too far from the grid). In combination with other open source data such as Worldpop [10] or Global Human Settlement Layer [11], it could also provide rough population estimates for the detected villages. This would then enable decision makers to use algorithms that weigh parameters such as population, surface area and distance to the grid in order to prioritize villages according to these parameters.

References

1. IEA, *World Energy Outlook 2017* (OECD Publishing, Paris/IEA, Paris, 2017), p. 115
2. P. Bertheau, A. Oyewo, C. Cader, C. Breyer, P. Blechinger, Visualizing national electrification scenarios for Sub-Saharan African countries. Energies **10**(11), 1899 (2017)
3. D. Kaundinya, P. Balachandra, N. Ravindranath, Grid-connected versus stand-alone energy systems for decentralized power—a review of literature. Renew. Sustain. Energy Rev. **13**(8), 2041–2050 (2009)
4. Sentinel-2 Images. http://www.esa.int/Our_Activities/Observing_the_Earth/Copernicus/Sentinel-2/Data_products. Accessed 21 Mar 2018
5. Level 2A Products. https://earth.esa.int/web/sentinel/user-guides/sentinel-2-msi/product-types/level-2a. Accessed 21 Mar 2018
6. T. Brinkhoff, City population—population statistics in maps and charts for cities, agglomerations and administrative divisions of all countries of the world [online]. (2018), http://citypopulation.de. Accessed 10 Oct 2017
7. L. Congedo, Semi-automatic classification plugin documentation. Release 5(3.6.1) (2016)
8. C. Song, C. Woodcock, K. Seto, M. Lenney, S. Macomber, Classification and change detection using Landsat TM data. Remote Sens. Environ. **75**(2), 230–244 (2001)
9. RapidEye Satellite Imagery. https://www.planet.com/products/satellite-imagery/files/160625-RapidEye%20Image-Product-Specifications.pdf. Accessed 21 Mar 2018
10. WorldPop Population Layer. http://www.worldpop.org.uk/data/get_data/. Accessed 21 Mar 2018
11. Global Human Settlement Layer. http://ghsl.jrc.ec.europa.eu/data.php. Accessed 21 Mar 2018

Chapter 8
Rural Household Electrification in Lesotho

M. Mpholo, M. Meyer-Renschhausen, R. I. Thamae, T. Molapo, L. Mokhuts'oane, B. M. Taele and L. Makhetha

Abstract Despite serious efforts of the Lesotho Government, Lesotho Electricity Company (LEC) and other stakeholders, the level of rural household electrification and affordability are still low. Whereas in 2015 about 72% of urban households were grid-connected, this was only true for 5.5% of rural households. Furthermore, the vast majority of rural households use fuel wood, while electricity use, where available, represents a small share of the domestic energy consumption. The LEC data shows that the average consumption per household has decreased by over 60% between 2001 and 2016 in urban households. This indicates that the bulk of new connections are to the rural poor households. This is plausible given that majority of households perceive electricity and other commercial sources of energy to be more

---

M. Mpholo (✉) · T. Molapo · B. M. Taele
Energy Research Centre and Department of Physics and Electronics, National University of Lesotho, Roma 180, Lesotho
e-mail: mi.mpholo@nul.ls

T. Molapo
e-mail: td.molapo@nul.ls

B. M. Taele
e-mail: bm.taele@nul.ls

M. Meyer-Renschhausen
Hochschule Darmstadt University of Applied Sciences, Haardtring 100, 64295 Darmstadt, Germany
e-mail: martin.meyer-renschhausen@h-da.de

R. I. Thamae
Energy Research Centre and Department of Economics, National University of Lesotho, Roma 180, Lesotho
e-mail: r.thamae@nul.ls

L. Mokhuts'oane
Department of Energy, Rural Electrification Unit, Maseru 100, Lesotho
e-mail: lmokhutsoane@yahoo.com

L. Makhetha
Department of Economics, National University of Lesotho, Roma 180, Lesotho
e-mail: lesekomakhetha@gmail.com

M. Mpholo et al. (eds.), *Africa-EU Renewable Energy Research and Innovation Symposium 2018 (RERIS 2018)*, Springer Proceedings in Energy,
https://doi.org/10.1007/978-3-319-93438-9_8

expensive than the traditional biomass. Therefore, the paper discusses this existing status quo with regard to rural electrification using data from the major players such as LEC, Rural Electrification Unit and Bureau of Statistics.

Keywords Rural electrification · Affordability · Lesotho

8.1 Introduction

Rural electrification has received a substantial attention from policy-makers, donors and scholars. Its policies, which have been changing overtime, have been initiated strongly by the World Bank on the premise that rural electrification acts as a catalyst for rural development [1]. As a result, countries were encouraged to liberalise their energy markets, introduce transparent forms of regulations and attract private investment. Currently, there are special initiatives by the World Bank with regard to rural electrification in Africa. Firstly, the World Bank helps in addressing issues of serving dispersed rural energy demand and reforming power sectors. These include finding solutions for improving energy access for the excluded population and development of local delivery and financing mechanisms. Secondly, the World Bank facilitates access to cooking fuels through programs emphasising fuel-wood management, stove efficiency, charcoal efficiency and transition to modern fuels. Finally, it assists in policy reforms including opening markets, elimination of price distortions and facilitating entry of competitors in order to improve rural energy issues [2].

Like many other African countries, Lesotho has established rural electrification programmes that aim at bringing electrical power to rural and remote areas. Despite the serious efforts of the Government of Lesotho, Lesotho Electricity Company (LEC) and other stakeholders, the level of households connected to the grid is still low. In 2015, about 72% of the households in urban regions were connected to the grid while the share of households in the rural areas with access to electricity was just 5.5% [3]. Even in regions covered by the grid, not all households are connected to the grid. Furthermore, households connected to the grid are often reluctant to make use of electricity intensively, as evidenced by the declining average household consumption depicted in Fig. 8.1. As a result, a significant share of energy services is still produced by other forms of energy like traditional biomass, paraffin or diesel. Hence, if a significant share of the households connected to the grid uses it for a few purposes only, economies of scale are not fully exploited. This is evidenced by Fig. 8.1 which shows how the LEC customer base has increased by almost a factor of 10 from around 25,000 in 2001/02 to approaching 210,000 in 2016/17 although the average consumption per household decreased by over 60% during the same period (2951–1157 kWh/year). As a consequence, capital invested in the expansion of the power grid is flowing back very slowly. Thus, lacking grid coverage is not the only obstacle to a more comprehensive use of electricity in Lesotho.

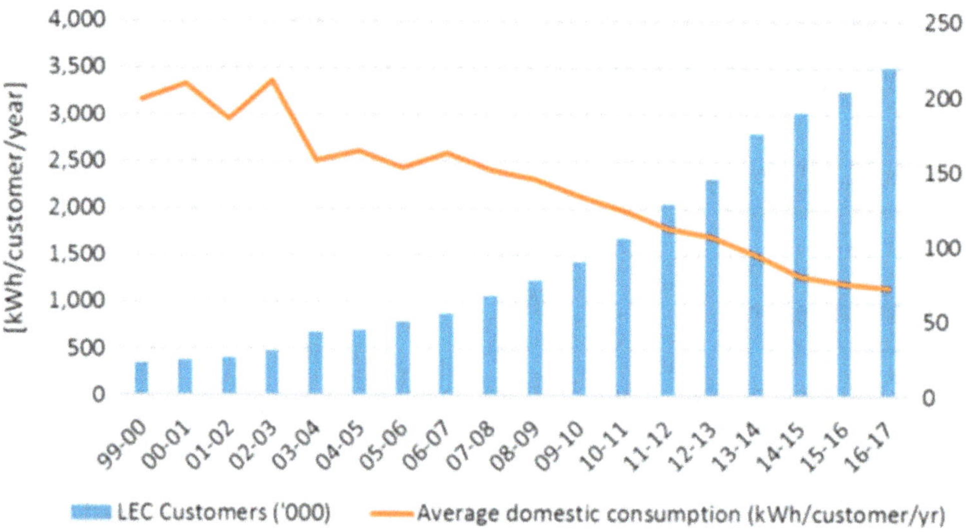

Fig. 8.1 LEC customer numbers and average consumption per domestic customer 2000–2016 [3]

The target of this paper is to discuss the existing status quo with regard to rural electrification using data from the major players such as LEC, Rural Electrification Unit (REU) and Bureau of Statistics (BoS). Whereas, data on the share of electrified households and regions are provided by official statistics, the national power provider and the regulator, solid information about the impediments for grid connection and substituting electricity for traditional energy needs is lacking. The information about these impediments is an essential prerequisite for a successful policy strategy and needs to be addressed by empirical studies.

The remainder of the paper is organized as follows. Section 8.2 includes a brief literature review on the global challenges and success factors of rural electrification. Section 8.3 discusses the status of rural electrification in Lesotho. The last section offers discussions and conclusions.

8.2 The Challenges and Success Factors of Rural Electrification

The major challenges to rural electrification, especially in developing countries, are significantly attributed to the capital costs and limited returns in the short and medium term. The investment for grid extension and off-grid schemes to reach remote and scattered households are often substantially high yet the low electricity consumption level of rural households, along with tariff policies meant to equalise the price of a kilowatt-hour between rural and urban areas, imply limited returns [4]. The World Bank also arguably postulates that the challenge with rural electrification is not necessarily with low electricity consumption but the fact that rural customers often cannot get affordable credit. As a result, this makes it hard for them to pay the high start-up costs of improving their energy supplies, which also has implications on utility companies' returns on supplying electricity to rural areas [2].

Considering macroeconomic success factors of rural electrification, Mostert emphasises the importance of two key factors—GDP per capita and the share of population living in rural areas [5]. "It seems that once a country passes two thresholds: a national electrification rate above 50% and a per capita income higher than US$3000 on a purchasing power parity (PPP) basis—it becomes financially feasible to implement a rural electrification program to achieve '100% rural electrification' within a few years" [5, p. 18]. These factors are certainly a challenge for most African countries to meet. Furthermore, Barnes investigated successful approaches of rural electrification in both developing as well as developed countries [6]. The conclusion is that there is not a single successful approach of rural electrification, but a variety of approaches can be successful as long as a high degree of autonomy "—in which the implementing agency can pursue rural electrification as its primary objective," is given [6, p. 344].

8.3 Current Electrification Situation in Lesotho

To address the rural electrification challenge in Lesotho, there are major interdependent issues that need to be analysed and attended to which include the following: the mountainous terrain; conducive policy environment; and change of attitudes towards cleaner energy. First, Lesotho is a mountainous country with a rough terrain and the altitude varies from 1500 to 3482 m. This geographic situation has impact on the population density and hence on access to electricity services. For example, 56% of the population is concentrated in the lowlands which cover only 17% of the total area [7]. The remaining large area covers mountain and foothills regions which are mainly rural and characterised by scattered settlement patterns. As a result, the terrain is a challenge for national grid electrification and the scattered settlements present low numbers for economic returns on the needed capital infrastructure [8]. For example, the national household electrification rate was around 35% in 2015 with the urban rate estimated at 72% while the rural was 5.5% [3].

Second, the Lesotho Government established the Rural Electrification Unit (REU) in 2004 in order to extend the grid to rural areas not serviced by the national power utility, LEC. The REU is mandated to serve rural areas and other areas which are more than 3.5 km away from the LEC transmission and distribution lines. It normally responds to the requests made through organised village schemes. It has a backlog of electricity schemes seeking services growing at a high rate. For instance, new schemes, 135, registered with the REU in 2016/17 brought the list to a total of 680 schemes. Nevertheless, only 26 schemes were served during that financial year, whereby 6818 new households were electrified at a cost of 109 million Maloti which was financed by the Lesotho Government [9].

Over and above the LEC grid extension projects, REU has implemented three grid extension pilot access projects, owned and operated by it. These are in Qholaqhoe (Butha-Bothe), Mpiti-Sekake (Qacha's Nek) and Dilli-Dilli-Sixondo (Qacha's Nek) which is connected directly to the South African grid line. Since this is a post-payment system, REU has carried out a number of household due to non-payment. In Qholaqhoe, 58 customers (out of 308) have been disconnected which represents a 16% disconnection rate while in Dilli-Dilli-Sixondo, 86 customers out of 371, which represents 23%, were disconnected. On the other hand, the operations and maintenance of all the grid projects continue to run at a loss, as seen in Table 8.1, and the deficit is covered by the government [9].

Lastly, households in Lesotho, especially those that are in rural areas, use biomass (i.e. fuel wood, agricultural residues and dung) predominantly while other intermediate fuels (i.e. coal and kerosene) and modern fuels (i.e. electricity and LPG) represent a small share of the total domestic energy consumption, as depicted in Fig. 8.2. In areas where electricity is available, it is mainly used for lighting, TV and radios rather than for cooking and heating [10]. These occur because majority of households perceive electricity and other cleaner sources of energy such as gas, solar and batteries as more expensive than traditional biomass fuels [11]. Therefore, the perceived affordability of electricity by households, particularly in rural areas, seems to be one of the main barriers to using more of electricity and other cleaner energy sources.

8.4 Discussion and Conclusions

Rural electrification is a complex multi-pronged challenge as it touches on many inter-related issues. In Lesotho, the majority of the population still resides in rural areas where settlements are normally scattered and in geographically difficult to reach places due to the mountainous terrain. This makes it very costly to extend the national grid to such communities with no returns on capital investment as the consumption is bound to be very low due to higher rates of unemployment in rural areas which result in electricity being unaffordable.

Table 8.1 Financial operation of the REU pilot projects [9]

Project	Bulk power purchased (M)	O&M costs	Electricity sales	Deficit (M)
Qholaqhoe/ Makhunoane	379,810.27	60,000.00	186,634.38	−253,175.89
Dilli-Dilli/ Sinxondo	236,361.50	63,600.00	215,412.83	−84,548.67
Mpiti-Sekake	1,032,085.06	96,360.00	877,422.90	−251,002.16
Total	1,648,256.83	219,960.00	1,279,490.11	−588,726.72

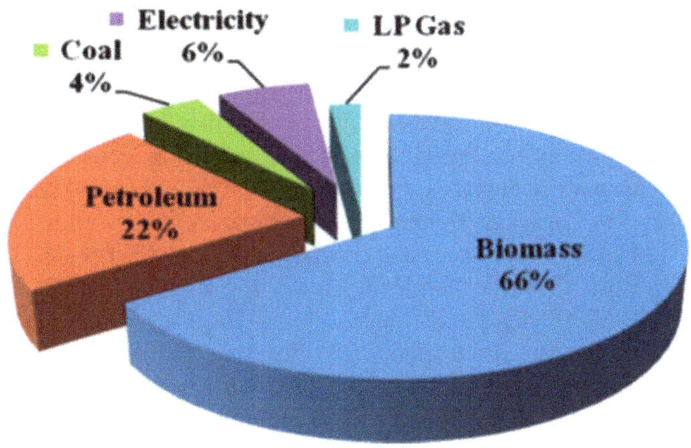

Fig. 8.2 Lesotho's energy consumption in 2008 [10]

On average, households connected to the grid do not use electricity in a comprehensive way. As a consequence the average consumption of grid connected households tends to drop and the utility's economic performance deteriorates. This is a general challenge of all developing countries that invest in increasing the rural electrification rate. Demand normally matures slowly (over 2–3 years and even later) as consumers wire their houses, invest in appliances and make a switch from other fuels for lighting, cooking and heating. Unfortunately, this progression is difficult to predict, making returns on investment in grid extension to poor rural people uncertain [6].

One effective solution could be to consider utilisation of Renewable Energy Technologies mini-grids instead of using the funds to extend the national grid. In this manner once the capital infrastructure has been installed, the only costs incurred are those of operations and maintenance that are less significant, which then could translate in low tariffs to the connected rural poor resulting in affordability, hence higher consumption.

Acknowledgements The following organisations are acknowledged for financially supporting this work: The Elsevier Foundation—TWAS Sustainability Visiting Expert Programme and Materials Research Science and Engineering Center of the University of Pennsylvania (Grant No. DMR-1120901).

References

1. World Bank, *The Welfare Impact of Rural Electrification: A Reassessment of the Costs and Benefits; An IEG Impact Evaluation* (World Bank Publications, 2008)
2. World Bank, *Rural Energy and Development for Two Billion People: Meeting the Challenge (English)* (World Bank, Washington D.C., 2005)
3. Lesotho Electricity and Water Authority, *Electricity Supply Cost of Service Study: Load Forecast Report* (LEWA, Maseru, 2017)
4. T. Bernard, Impact analysis of rural electrification projects in sub-Saharan Africa. World Bank Res. Observer **27**(1), 33–51 (2010)

5. W. Mostert, *Review of Experiences with Rural Electrification Agencies: Lessons for Africa* (European Union Energy Initiative Partnership Dialogue Facility, Eschborn, 2008)
6. D.F. Barnes, Meeting the challenge of rural electrification in developing nations: the experience of successful programs, in *Energy Sector Management Assistance Program* (World Bank, Washington D.C., 2005), Conference Version, March, pp. 1–16
7. Bureau of Statistics, *Lesotho Demographic Survey Tables*, vol. III (BOS, Maseru, 2011)
8. B.M. Taele, L. Mokhutsoane, I. Hapazari, An overview of small hydropower development in Lesotho: challenges and prospects. Renew. Energy **44**, 448–452 (2012)
9. Rural Electrification Unit, *Annual Report for 2016/17* (REU, Maseru, 2017)
10. B.M. Taele, L. Mokhutsoane, I. Hapazari, S.B. Tlali, M. Senatla, Grid electrification challenges, photovoltaic electrification progress and energy sustainability in Lesotho. Renew. Sustain. Energy Rev. **16**, 973–980 (2012)
11. Deutsche Gesellschaft für Technische Zusammenarbeit (GTZ), *Lesotho Energy Access Strategy Project: Baseline Study Report* (GTZ, Maseru, 2007)

Chapter 9
Simulation and Optimization of Renewable Energy Hybrid Power System for Semonkong, Lesotho

Leboli Z. Thamae

Abstract Rugged hills and mountain ranges with sparsely populated rural villages characterize the vast majority of Lesotho's landscape, making it prohibitively expensive and financially unviable to connect these remote villages to the national electricity grid. This lack of access to electricity has hampered many social and economic developments due to insufficient provision of much-needed power to homes, schools, police stations, clinics and local businesses. This paper proposes a renewable energy hybrid power generation system for one such remote town of Semonkong, in Maseru district, Lesotho. The study models, simulates and optimizes the hybrid power system using the load profile of Semonkong town and the available renewable resources data of solar radiation, wind speeds and water flow rates from the nearby Maletsunyane River. The HOMER software is used to provide an optimal system configuration in terms of the minimum levelized cost of electricity (LCOE) and the maximum renewable energy fraction, based on various renewable and alternative energy sources of solar photovoltaic, wind turbine, mini-hydro turbine, diesel generator and battery storage. Sensitivity analysis on solar radiation, wind speed, stream flow, diesel price and energy demand is undertaken to evaluate the feasibility of a completely-renewable power system suitable for this remote area application. Simulation results for the isolated optimized hydro/wind/PV/diesel/battery hybrid system configuration achieves LCOE of US$0.289/kW at a renewable energy fraction of 0.98. Thus, the diesel generator will always be required to augment power supply for Semonkong especially during the dry and cold winter months of May to September when the energy demand is at its peak but the solar radiation and stream flow are at their lowest.

Keywords Renewable energy · Solar photovoltaic · Wind energy
Mini-hydropower · Optimal hybrid system

L. Z. Thamae (✉)
Department of Physics and Electronics, National University of Lesotho, Rome 180, Lesotho
e-mail: zl.thamae@nul.ls; thamae@gmail.com

L. Z. Thamae
Energy Research Centre (ERC), National University of Lesotho, Rome 180, Lesotho

M. Mpholo et al. (eds.), *Africa-EU Renewable Energy Research and Innovation Symposium 2018 (RERIS 2018)*, Springer Proceedings in Energy,
https://doi.org/10.1007/978-3-319-93438-9_9

9.1 Introduction

It has been estimated that around 600 million people in Sub-Saharan Africa lived without access to electricity at all [1]. In Lesotho for instance, around 60% of the population, mostly in isolated rural areas, currently do not have access to electricity. Extension of the grid to such areas is highly costly and often not feasible, at least in the short to medium term [2]. In such cases, mini-grids based mainly on locally-available renewable resources for decentralized hybrid power systems and household energy solutions become handy in provision of electricity for households and local businesses. Hybrid mini-grids are a mature and reliable solution that combines at least two different kinds of technologies for power generation to be used for lighting, communications, water supply or motive power [3]. A small town of Semonkong, located in Lesotho's Maseru district, is one such rural village that is currently supplied with a hydro-diesel hybrid power system using a village-wide distribution network, which is at least 70 km away from the nearest grid. Due to the challenging mountainous terrain and difficult access for transmission lines, grid extension to Semonkong will come at a very high cost.

Semonkong power station's mini-hydro component is a run-of-river system with a small dam/reservoir built on the Maletsunyane River that passes through the village. It uses a vertical Francis turbine rated at 180 kW with a head of 19 m. The backup diesel generator is rated at 500 kVA (or 400 kW at 0.8 power factor). The peak electricity demand in August 2017 has been recorded as 198 kW with an average energy demand of 3.6 MWh/day. For seasons with good rains in the past, the hydro turbine generation was able to meet the local electricity demand, even though in recent years the growing peak demand at around 200 kW has exceeded the mini-hydro turbine capacity. During the 2012/13, the annual energy production for Semonkong power station amounted to about 505 MWh (or 1.383 MWh/day), with 89% coming from hydro and 11% from diesel, the latter requiring an actual cost of M0.557 million [4], which translates to M9.84/kWh, equivalent to US$0.70/kWh.[1] Comparing this value with the approved electricity tariffs during the same year, the unit tariffs for general purpose and domestic customers were M1.01/kWh (or US$0.07/kWh) and M0.90/kWh (or US$0.06/kWh) respectively [4]. It can therefore be deduced that, excluding labour and operation and maintenance (O&M) costs for the isolated Semonkong mini-grid, the diesel-generated unit of energy was on average 10 times more expensive than its selling price. Since electricity tariffs are the same for these rural customers as they are for the grid-connected ones despite dramatically different costs, this has cross-subsidy implications and increases the overall price for all consumers in the country.

Another known challenge of diesel generator stations is that it is difficult to distribute the fuel especially in rural and remote mountainous areas. It is against this background that 100% diesel-fuelled mini-grids will prove to be more expensive on a lifetime basis than hybrid ones, strengthening the case for off-grid electrification

[1]Assuming US$1.00 = M 14.00 (LSL) in 2012/13.

using a combination of locally-available renewable resources of solar, wind and hydro where feasible, with diesel generator and battery bank as back-up components [5]. This study therefore seeks to evaluate the type of renewable energy technologies (RETs) that are mostly favourable for the study area of Semonkong town, in Lesotho. It will answer research questions such as whether it is feasible and cost-effective to supply the town completely from renewable energy; whether it is critical to add battery bank and diesel generator back-up systems; and whether the hybrid power system design will meet the growing electric demand. The data obtained for the load profiles, renewable and alternative resources, and components costs are briefly described under the methodology in Sect. 9.2, followed by the results and discussions of the proposed full hybrid PV/wind/hydro/diesel/battery system configuration, simulation, optimization and sensitivity analysis in Sect. 9.3. The conclusions are summarized in Sect. 9.4.

9.2 Methodology: Load Profile, Resources and Components

The main aim of the study is to use HOMER software to evaluate hybrid power system designs for Semonkong using several available RETs incorporating diesel generator and battery back-up systems. The approach engaged in answering the above research questions makes use of HOMER software (which stands for Hybrid Optimization Model for Electric Renewable) [6–8]. As inputs to HOMER, data on electricity load profile and resource availability for the rural town of Semonkong in Lesotho has been collected from several reliable sources [4, 9]. Component costs and performance characteristics have been benchmarked with similar previous studies [3, 5, 7, 8, and 10]. Semonkong is a good case study for off-grid hybrid power system due to one river (Maletsunyane) running through it and having meteorological stations with data available for river flow rates, solar radiation and wind speeds. Its location therefore offers excellent natural conditions for the use of mini-hydro power, solar PV and wind as illustrated in the following sub-sections.

9.2.1 Load Profile

A load profile for Semonkong is demonstrated in Fig. 9.1 based on the utility's 2016/17 electricity demand data [9], entered in HOMER with a random variability of 5%. The hourly load consumption shifts throughout the day with peak demand of 212 kW, average load of 150 kW, average energy consumption of 3611 kWh/day and a load factor of 70.9%.

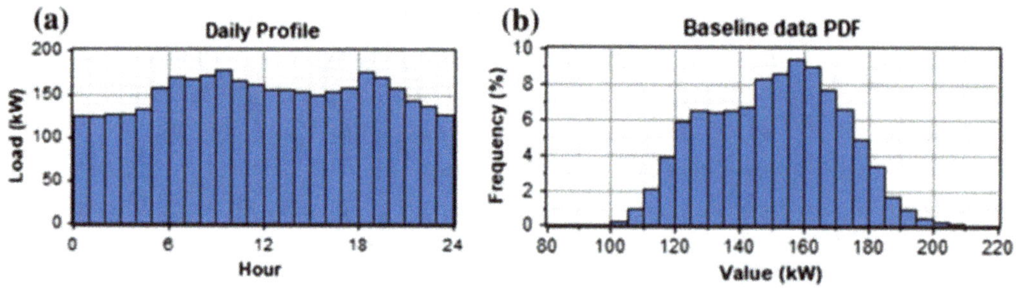

Fig. 9.1 Semonkong's load profile: **a** Daily profile. **b** Demand PDF

9.2.2 Resources and Components

Figures 9.2, 9.3 and 9.4 graphically illustrate the respective data for renewable resources of river flow rates, solar radiation and wind speeds at the study area. According to Fig. 9.2, there is more water available during the rainy summer months of October to April than during the dry winter months of May to September. The stream flow achieves an annual average of 1095 L/s. Figure 9.3 indicates daily solar radiation with annual average global horizontal radiation of 5.43 kWh/m^2/d and average clearness index of 0.61. Similar to the hydro resource, the insolation is higher during the summer months but slightly weaker (about 4 kWh/m^2/d or lower) in dry winter months of May to July while the clearness index is higher. The monthly average wind speed illustrated in Fig. 9.4 is particularly stronger between August and October and it has a Weibull probability distribution function (PDF) with a shape factor of k = 2.03. The annual average wind speed is about 6.85 m/s, which is higher than the 4–5 m/s normally needed to make a wind power system profitable [2].

To enter the components data for HOMER simulation, the existing 180 kW hydro turbine is modelled with component cost of US$1790/kW for the initial capital cost, replacement cost that is 7% lower, O&M costs at 2% of capital cost for a lifetime of 25 years. A solar PV component cost of US$2822/kW is used for the capital cost, US$2500/kW for replacement and 1.5% of initial capital for O&M costs with a projected lifetime of 20 years. Fuhrlander FL30 wind turbine with a rated power of 30 kW at 12 m/s and hub height of 27 m is used in HOMER with component cost of US$2120/kW for both the initial capital and the replacement costs. The O&M cost is taken to be 10% of the initial cost and the projected lifetime is 25 years. The 400 kW rating of the existing diesel generator is used with US $400/kW for both capital and replacement costs, US$1.75/hr for O&M costs and US$0.75/L diesel price in Lesotho as at November 2017. Battery voltage rating of 6 V with a nominal capacity of 360 Ah (2.16 kWh) has been chosen for simulation with 4 batteries per string at an initial unit cost and replacement cost of US$225 and O&M cost of US$10. A converter for DC to AC power conversion has also been included at US$1445/kW capital and replacement costs and 10% O&M costs.

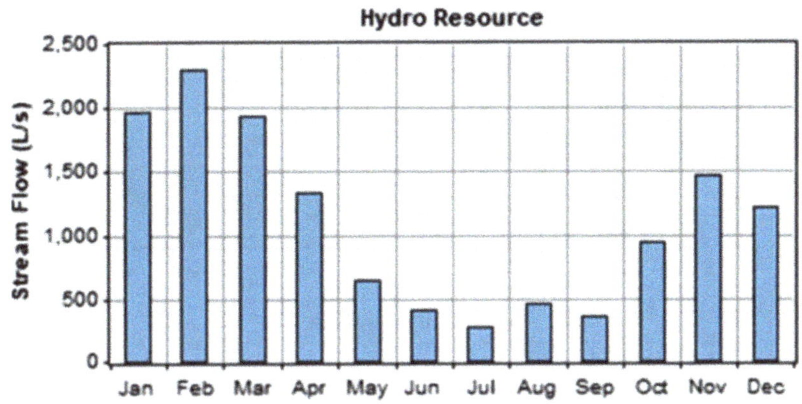

Fig. 9.2 Maletsunyane River's average flow rates

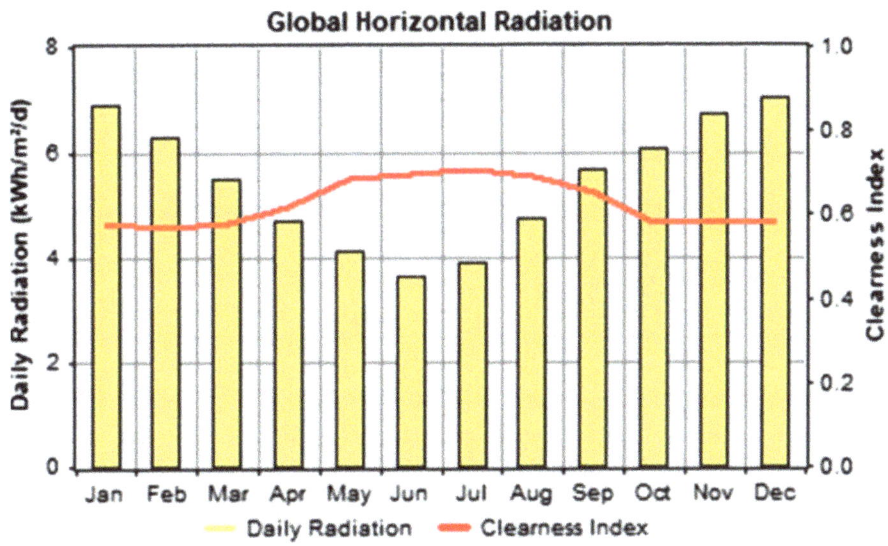

Fig. 9.3 Semonkong's daily solar radiation and clearness index

9.3 Results and Discussions: Hybrid System Simulation and Optimization

A hybrid power generation system using the local load profile, resources and components costs from Sect. 2 has been modelled in HOMER. Since the load profile and natural conditions are site specific, the results obtained in the next sub-sections are valid for the study site of Semonkong and should not be extrapolated to other areas. However, the lessons learnt will be applicable to locations with similar settings.

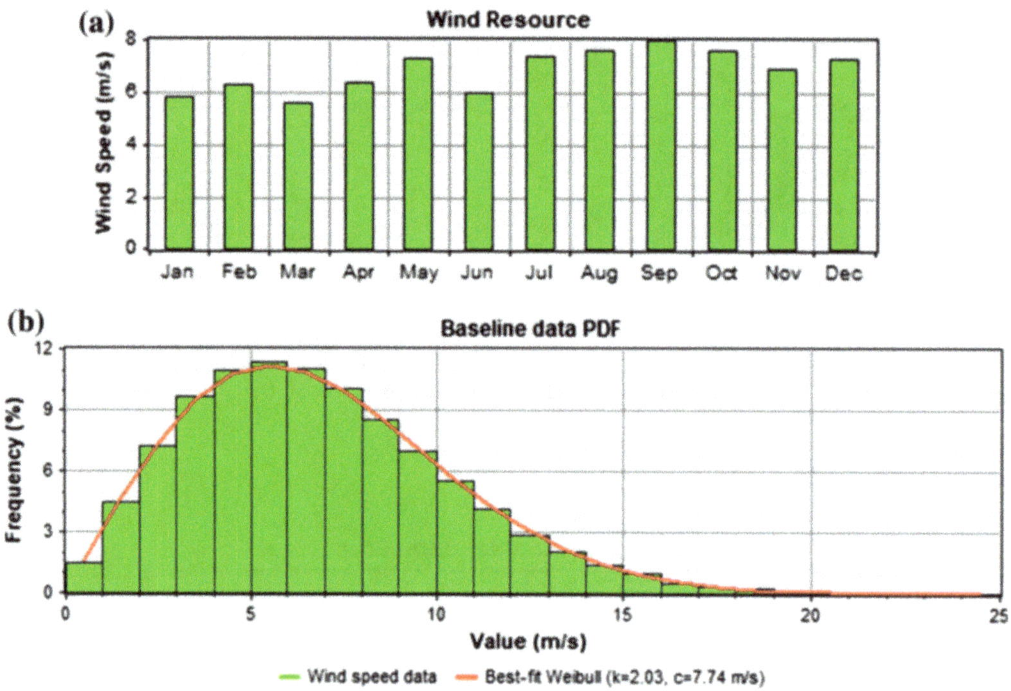

Fig. 9.4 Semonkong's wind resources: **a** Monthly average wind speeds. **b** Wind speed PDF

9.3.1 Simulation Results

HOMER simulation produces a list of feasible design options or system configurations, ranked by the lowest total net present cost (NPC) as illustrated in Fig. 9.5. Examining each feasible system configuration allows for economic and technical merits evaluation including the LCOE and renewable energy fraction. Other system operational characteristics such as annual power production, annual electric load served, fuel consumption, excess electricity, capacity shortage, unmet electric load and emissions can also be evaluated.

For instance, the simulation results from Fig. 9.5 show that the existing hydro/diesel hybrid system is ranked bottom as one of the least cost-effective options at US$0.471/kWh (M6.594/kWh) and 0.66 renewable energy fraction (66% hydro and 34% diesel). The hydro component is more effective during the rainy summer season but needs diesel generator support during the dry winter months (compare with Fig. 9.2). This configuration has 26.4% excess electricity production per annum. The share of renewables needs to be increased to make this hybrid system almost independent of diesel and to lower energy prices over the long term as described in the next sub-section.

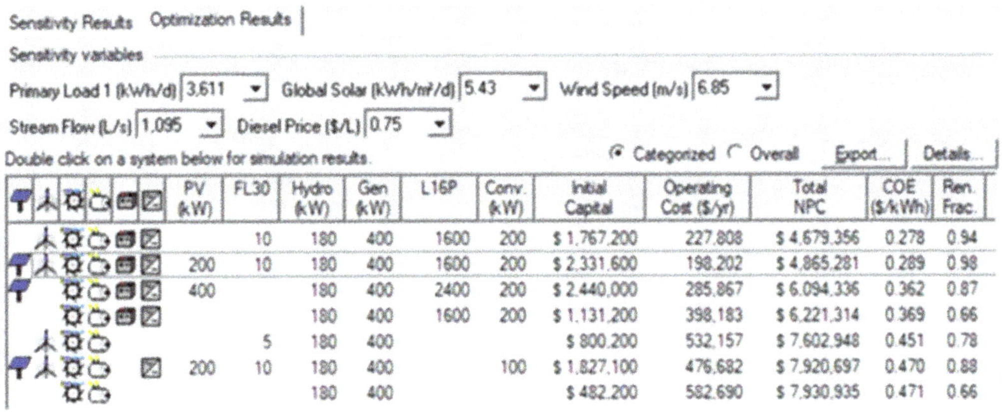

Sensitivity Results Optimization Results

Sensitivity variables

Primary Load 1 (kWh/d) 3,611 Global Solar (kWh/m²/d) 5.43 Wind Speed (m/s) 6.85

Stream Flow (L/s) 1,095 Diesel Price ($/L) 0.75

Double click on a system below for simulation results. Categorized Overall Export... Details...

PV (kW)	FL30	Hydro (kW)	Gen (kW)	L16P	Conv. (kW)	Initial Capital	Operating Cost ($/yr)	Total NPC	COE ($/kWh)	Ren. Frac.
	10	180	400	1600	200	$ 1,767,200	227,808	$ 4,679,356	0.278	0.94
200	10	180	400	1600	200	$ 2,331,600	198,202	$ 4,865,281	0.289	0.98
400		180	400	2400	200	$ 2,440,000	285,867	$ 6,094,336	0.362	0.87
		180	400	1600	200	$ 1,131,200	398,183	$ 6,221,314	0.369	0.66
	5	180	400			$ 800,200	532,157	$ 7,602,948	0.451	0.78
200	10	180	400		100	$ 1,827,100	476,682	$ 7,920,697	0.470	0.88
		180	400			$ 482,200	582,690	$ 7,930,935	0.471	0.66

Fig. 9.5 Hybrid power system simulation results

9.3.2 *Optimization Results*

Using the simulation results of Fig. 9.5, the wind/hydro/diesel/battery combination that excludes the solar PV component is the most cost-effective mini-grid solution for Semonkong with LCOE of US$0.278/kWh (M3.892/kWh) and a renewable energy fraction of 0.94. This optimum system architecture will have a total annual energy production of 44% wind, 50% hydro and 6% diesel, with 41.5% excess annual electricity production. Though the unit cost would be about half compared to the hydro/diesel hybrid above, it will still be more expensive than the current utility's power purchases from the local 'Muela hydro at a subsidized M0.15/kWh (US$0.011/kWh), imports from South Africa's Eskom at an average of M0.97/kWh (US$0.069/kWh) and Mozambique's EDM at around M1.50/kWh (US$0.107/kWh). Even the current 2017/18 respective domestic and general purpose approved tariffs of M1.424/kWh (US$0.102/kWh) and M1.6/kWh (US$0.114/kWh) will be less than half of the US$0.278/kWh LCOE [4].

The second optimal configuration from Fig. 9.5 is the PV/wind/hydro/diesel/battery hybrid with LCOE of US$0.289/kWh (M4.046/kWh) but marginally favourable renewable energy fraction of 0.98. Figure 9.6 shows its system architecture together with electrical operational characteristics. The annual electric energy production comprises 13% PV, 39% wind, 45% hydro and 2% diesel with an excess annual generation of 48.3%.

One clear deduction from these two most cost-effective results is that it will not be possible to supply Semonkong on RETs only as the diesel generator is always required in any system configuration to meet the local demand. Diesel generator use is critical for quality of service when other RETs are low and unable to meet the demand in winter months of June to September. However, it can be minimized from 6 to 2% by employing more renewables as in the second option that makes use of all the available natural renewable resources, with half the pollutant emissions but small difference of 3.8% in LCOE. Hence for the sensitivity analysis in the

Fig. 9.6 Optimized PV/wind/hydro/diesel/battery hybrid

following sub-section, the two most cost-effective system architectures with and without the PV component are considered.

9.3.3 *Sensitivity Analysis Results*

The sensitivity analysis results for the optimal system with fixed primary load, stream flow and diesel price superimposed with LCOE is illustrated Fig. 9.7. It can be seen that the average local natural conditions of 5.43 kWh/m^2/d solar radiation and 6.85 m/s wind speed lead to LCOE of US\$0.278/kWh for a hydro/wind/diesel/battery system configuration indicated by a diamond. This configuration remains dominant regardless of changes in solar radiation and wind speeds. However, for wind speed, the LCOE varies from US\$0.299/kWh at 6.0 m/s to a favourable US \$0.250/kWh for wind speeds around 8.0 m/s, regardless of the solar radiation.

Similar sensitivity analysis for the current load (3611 kWh/d) with decreased stream flow (900 L/s) and increased fuel price (US\$0.95/L), illustrated in Fig. 9.8 gives the LCOE of about US\$0.285/kWh for a hydro/wind/diesel/battery system architecture at the average local natural conditions of 5.43 kWh/m^2/d solar radiation and 6.85 m/s wind speed. This architecture becomes the most optimal combination for majority of the conditions, but when the solar radiation goes above 5.6 kWh/m^2/d for wind speeds between 6.0 and 6.3 m/s, the hydro/wind/PV/diesel/battery hybrid becomes more cost-effective. For varying solar radiation and wind speeds, the LCOE varies from a maximum of US\$0.311/kWh when they are respectively at 5.0 kWh/m^2/d and 6.0 m/s to a minimum of US\$0.255/kWh at wind speed of 8.0 m/s.

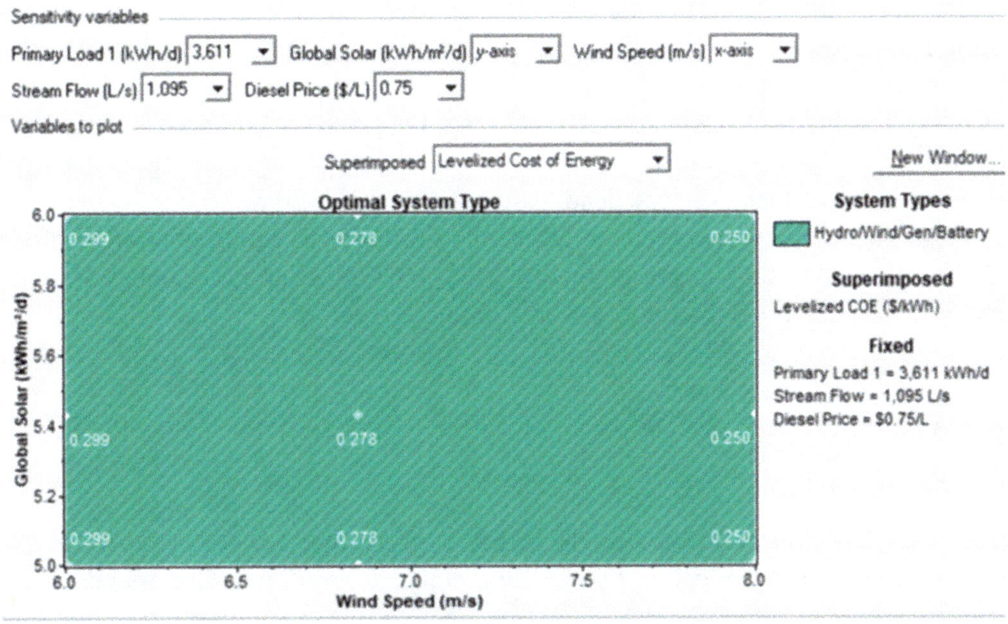

Fig. 9.7 Optimal system with fixed load, stream flow and diesel price

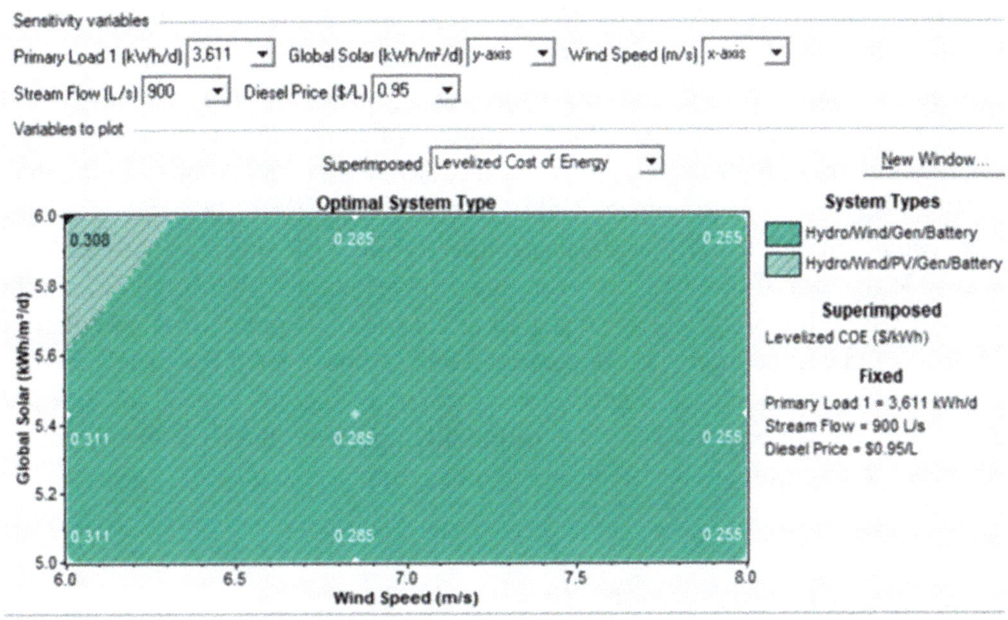

Fig. 9.8 Optimal system with fixed load, decreased stream flow and increased diesel price

The last sensitivity analysis assumes the local demand increase of 30% to 3960 kWh/d while the stream flow is still 900 L/s with diesel price at US$0.95/L. The average local conditions of solar radiation at 5.43 kWh/m^2/d and wind speed at 6.85 m/s lead to LCOE of about US$0.273/kWh for a hydro/wind/diesel/battery system architecture with slightly reduced renewable fraction of 0.92 and 33.6%

excess electricity. Hence the increase in local demand will still be met and at a slight 1.8% reduction in LCOE due to scale, but with the diesel generator's production increases from 6 to 8% to meet the increased new average demand of 195 kW. The hybrid system architecture integrating the PV component can play a negligible role for solar radiation values above 5.8 kWh/m^2/d as long as the wind speeds are lower than 6.1 m/s, leading to LCOE of US$0.303/kWh. Other solar radiation values and wind speeds at different values lead to maximum LCOE of US $0.304/kWh when they are respectively at 5.0 kWh/m^2/d and 6.0 m/s and a minimum of US$0.246/kWh for 6.0 kWh/m^2/d and 8.0 m/s.

9.4 Conclusions

This study has illustrated through HOMER hybrid power system simulation and optimization that technically, hybrid electricity generation using mainly local renewable energy resources can be a cost-effective means of supplying affordable and reliable power for rural communities like Semonkong in Lesotho. When compared with the existing hydro/diesel hybrid at the LCOE of US$0.471/kWh and 66% renewable fraction, a full hydro/wind/PV/diesel/battery hybrid system configuration could lower the costs by 40% to US$0.289/kWh with a higher renewable energy fraction of 98%. Clearly, the diesel generator backup will always be required to ensure reliability and quality of supply when the RETs are low.

The sensitivity analysis results further indicated that due to varying local natural conditions of solar, hydro and wind at the site, LCOE can vary from a minimum of US$0.250/kWh to a maximum of US$0.311/kWh. Increase in demand of around 30% would still be met by the chosen optimal hybrid system at a slightly reduced renewable fraction of 0.92. The actual required investment costs for upgrade of the already existing hydro/diesel hybrid station to the full PV/wind/hydro/diesel/battery hybrid system will be mainly on integrating wind turbines, solar PV panels, battery and converter components. Since this particular system is already owned and operated by the national utility, if the grid finally arrives in future, it would just be connected to the grid and augment the national supply.

References

1. A. Eberhard, K. Gratwick, E. Morella, P. Antmann, *Independent Power Projects in Sub-Saharan Africa; Lessons from Five Key Countries* (World Bank Group, Washington, D. C, 2016)
2. A.R.E. Report, *Hybrid Mini-Grids for Rural Electrification—Lessons Learned* (Alliance for Rural Electrification (ARE), Brussels, 2014)
3. S. Baurzhan, G.P. Jenkins, Off-grid solar PV: is it an affordable or appropriate solution for rural electrification in Sub-Saharan African countries? Renew. Sustain. Energy Rev. **60**, 1405–1418 (2016)

4. LEWA, Tariff Determination for LEC (Lesotho Electricity and Water Authority, 2018), http://www.lewa.org.ls/tariffs/Tariffs_Determinations.php. Accessed 02/03/2018
5. W. Su, Z. Yuan, M.-Y. Chow, Microgrid Planning and Operation: Solar Energy and Wind Energy, in *IEEE Power and Energy Society General Meeting*, 2010
6. NREL, Getting Started Guide for HOMER Legacy (Version 2.68) (National Renewable Energy Laboratory, Colorado, 2011)
7. D.K. Lal, B.B. Dash, A.K. Akella, Optimization of PV/Wind/Micro-Hydro/Diesel hybrid power system in HOMER for the study area. Int. J. Electr. Eng. Inf. **3**, 307–325 (2011)
8. T. Lambert, P. Gilman, P. Lilienthal, *Micropower System Modeling with HOMER, Integration of Alternative Sources of Energy* (Wiley, Hoboken, 2006)
9. Lesotho Electricity Company, www.lec.co.ls. Accessed 02/03/2018
10. U.S. Magarappanavar, S. Koti, Optimization of Wind-Solar-Diesel Generator Hybrid Power System using HOMER, in *International Research Journal of Engineering and Technology*, vol. 3(6), 2016

Chapter 10
Storage as the Weak Link of the Biomass Supply Chain

S. Dumfort, C. Kirchmair, K. Floerl, C. Larch and M. Rupprich

Abstract Biomass such as wood, straw or agricultural wastes are a worldwide abundant resource offering high potential for a decentralized energy production and supply which is especially interesting for rural areas. However, substance and energy loss caused by microbial degradation is one major reason for high feedstock costs. As a consequence of the microbial activity and further exothermic processes, heat is produced inside stored piles, leading to temperatures >200 °C, causing partial pyrolysis and self-ignition. This work investigates the degradation process of spruce forest residues in order to generate a better knowledge about underlying processes and possible counteractions. Therefore, the microbial metabolic activity has been described in dependency on the moisture content (MC), particle size distribution and pH by respirometric tests. Respirometric tests revealed the microbial activity over time showing a maximum within the first few days of storage. Our results show that the moisture content is a key factor during wood degradation. No microbial activity could be verified for a MC < 20%. A moisture content of 46% led to a monthly dry matter loss of 5.4%. Raising the pH to an alkaline environment reduced the monthly dry matter loss from 3.1 to 1.8% per month proofing it´s influence on microbial metabolic activity. Further investigations have to be conducted to clarify underlying mechanism and countermeasures.

S. Dumfort · M. Rupprich (✉)
Department of Environmental, Process & Energy Engineering,
MCI-The Entrepreneurial School®, 6020 Innsbruck, Austria
e-mail: marco.rupprich@mci.edu

S. Dumfort
e-mail: sabrina.dumfort@mci.edu

C. Kirchmair · K. Floerl
Bioenergie Tirol, 6020 Innsbruck, Austria
e-mail: christian.kirchmair@maschinenring.at

K. Floerl
e-mail: klaus.floerl@maschinenring.at

C. Larch
Syneco-Tec GmbH, 6067 Absam, Austria
e-mail: christoph.larch@syneco-consulting.it

M. Mpholo et al. (eds.), *Africa-EU Renewable Energy Research and Innovation Symposium 2018 (RERIS 2018)*, Springer Proceedings in Energy,
https://doi.org/10.1007/978-3-319-93438-9_10

Keywords Biomass · Dry matter loss · Self-ignition · Storage Woodchips

10.1 Introduction

For the generation of thermal energy, wood is the most commonly used biogenic resource, offering various advantages including CO_2-neutral energy production. Especially in Europe, wood is an abundant resource applicable independently of season and weather. In district heating plants especially, forest biomass in the form of high-quality woodchips, as well as woody waste such as bark, sawdust or forest residues, is a common combustible material. After harvest, the biomass is usually stored over a period of time before utilization. During the storage, destruction and conversion processes take place inevitably. Different bacteria and mold fungi colonize the biomass, causing its degradation. These microbes originate from the natural microbial community present in the biogenic material as well as from atmospheric deposition [1]. *Basidiomycetes* are the main wood rotters due to their ability to degrade cellulose, hemicellulose and lignin. These fungi can overcome difficulties in wood decay such as limited nutrient accessibility and the presence of antibiotic compounds (essential oils) [2]. Moreover, they can grow over a wide temperature and pH range. The degradation of the wood components is carried out in aerobic conditions [3]. These degradation or respiration processes result in an economical relevant dry-matter loss and consequently in a net-energy-value reduction.

Data regarding the weight loss of woodchips differ widely, indicating fluctuations due to different test setups used, unseasonal conditions and biomass properties. Also, the method being used for quantifying the dry-matter loss exerts great influence. Several European research groups are focusing on determining the behaviour of woodchips during storage and postulated an annual biomass loss of 10–40% [4–16]. Buggeln suggested a rule of thumb of 1% dry-matter loss per month in outside storage, which is accurate for woodchips of high quality [9]. However, this value is underestimating the dry-matter loss of woodchips containing bark or forest residues.

Primary factors influencing the degradation rate are the temperature, water content and oxygen availability [11]. Also, pH and nutrient availability are the main microbial growth factors either promoting or inhibiting microbial growth. These factors are mainly influenced by the pile geometry, piling method, particle size, comminution method, storage season, storage location and the tree species [3, 6, 10, 12, 13].

As an example, woodchips containing sawdust and fine materials such as needles offer a greater surface to be attacked by microbes and cause limited air passage, leading to higher pile temperatures [6]. Pecenka et al. [10] showed a monthly dry-matter loss of 1.0–3.6% for poplar woodchips of varying particle sizes, showing higher dry-matter loss for smaller particles. The compaction of the pile leads to heat accumulation and affects the fungal growth negatively. The pile geometry influences rainwater entry as well as pile compaction and heat dissipation depending on the volume-to-surface-area ratio [13, 14]. All mentioned studies proved a significant correlation between pile temperature and dry-matter loss.

However, detailed knowledge about what factors influence the degradation process to what extent are still missing, and further investigations have to be carried out for a better understanding of the underlying processes.

A woodchip pile by itself is an inhomogeneous system consisting of different layers, offering varying growth conditions for microorganisms leading to difficulties in the dry matter loss determination [16].

Against this background, a method has been implemented in the laboratory to simulate different storage conditions and pile layers in order to describe the degradation process according to its dependency on ambient conditions. Generally, the biodegradability of organic compounds is tested by respirometric tests, where the carbon dioxide (CO_2) produced during the degradation process is quantified. This method offers an accurate determination of the degradation rate under certain ambient conditions and can visualize the microbial activity. In this study, respirometric tests have been conducted for forest residues from spruce, aiming to describe the degradation process. The influence of the moisture content, particle size distribution and pH on the dry-matter loss has been investigated.

10.2 Materials and Methods

10.2.1 Respirometric Tests

The biodegradability of spruce woodchips and forest residues was tested via a modification of ISO-14852, where the CO_2 produced during the degradation process was quantified [17]. Figure 10.1 shows the experimental setup. The reactors were filled with 500–1000 g fresh woodchips and were incubated at varying moisture contents, particle sizes and pH values.

All reactors were covered with glass wool and aluminium foil as insulation and to keep dark test conditions. In order to keep aerobic conditions, air was continuously pumped into the reactor at a flow rate of about 0.01 m^3/h. Before entering the reactor, the air was passed through an absorption bottle with soda lime to remove atmospheric CO_2 and a bottle with either water for tests with moist material or silica gel for tests with dry test material. CO_2 that was produced during the test was flushed through three absorption bottles with 0.5 M sodium hydroxide, forming sodium bicarbonate. The concentration of CO_2 was determined by means of a two-step titration with hydrochloride acid, phenolphthalein and methyl orange as indicators. For calculating the wood degradation (%) it has been assumed that mainly hemicellulose is attacked and degraded within this period of time. Thermal gravimetric tests confirmed this assumption (data not shown).

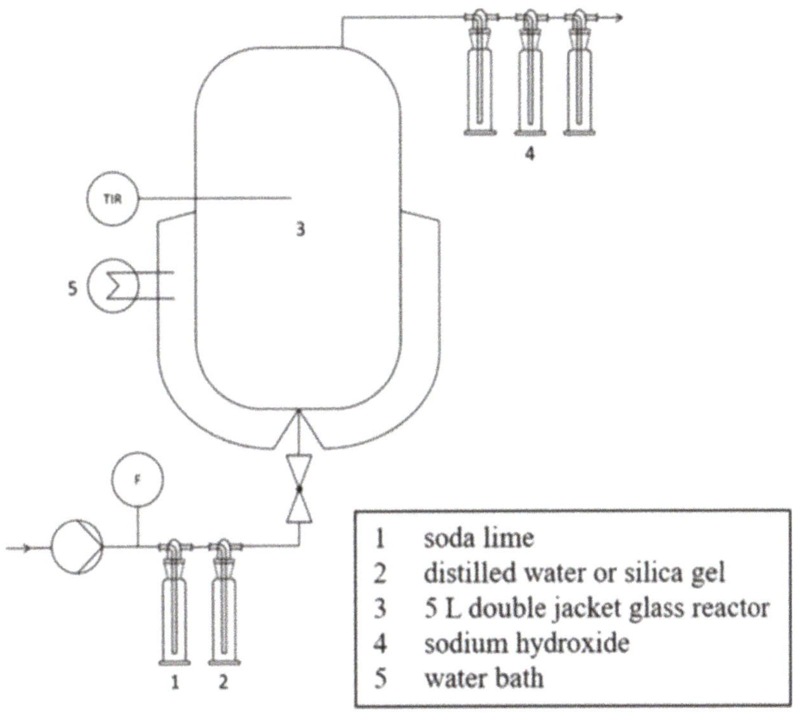

Fig. 10.1 Experimental setup of the respirometric test

10.2.2 *Samples and Test Conditions*

Prior to the test, fresh spruce woodchips were sieved according to ÖNORM EN 15149-1, and the reactors were filled with a defined quantity of each fraction [18]. The moisture content of the material was determined by drying three samples at 105 °C to constant weight at the beginning and at the end of each experiment.

In order to better understand the influence of the particle size distribution and the fine-material content on the degradation process, experiments were conducted with coarse, medium and fine woodchips according to Fig. 10.2. The coarse material contained a fine-material content (<3.15 mm) of 0%, medium 10% and fine woodchips 20%. A medium particle size distribution was selected for all other experiments.

The influence of the pH was investigated by adding 5 and 10% (w/w) of calcium carbonate to fresh forest residues and incubating for 45 days.

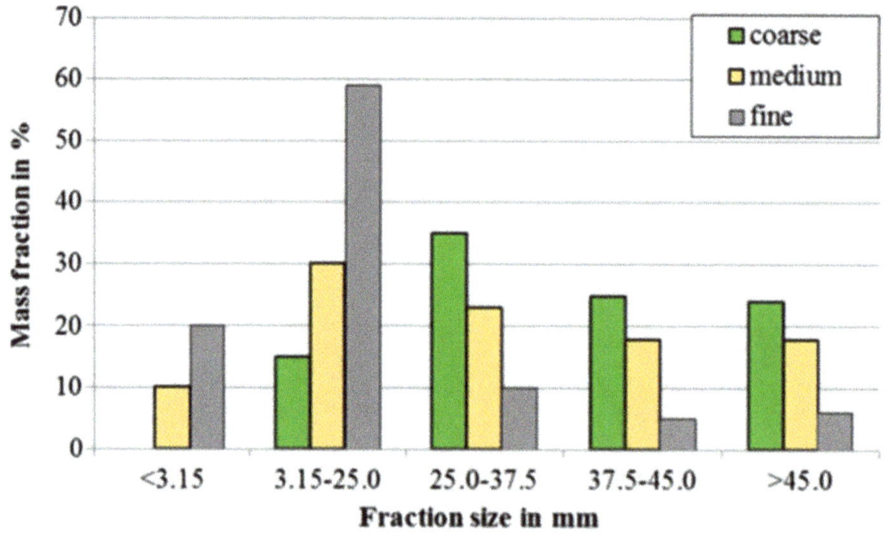

Fig. 10.2 Particle size distribution of fine, medium and coarse forest residues (n = 1)

10.3 Results and Discussion

10.3.1 Influence of the Moisture Content

Respirometric tests have been conducted with forest residues with a moisture content of 46, 30, 20 and 17%, respectively. The moisture content was lowered from 46% by means of a convection dryer at 20 °C. Figure 10.3 illustrates the CO_2 production rate over time and the dry matter losses at varying water contents.

Taking a look at the different curves, an increase of microbial activity with increasing moisture content is seen. Forest residues with a MC of 20% showed very low activity. Below this value, no activity could be proven, marking the minimum water availability needed for microbial growth. After 10 days the activity started to decrease for all test conditions. However, as our results showed, at 35% MC the

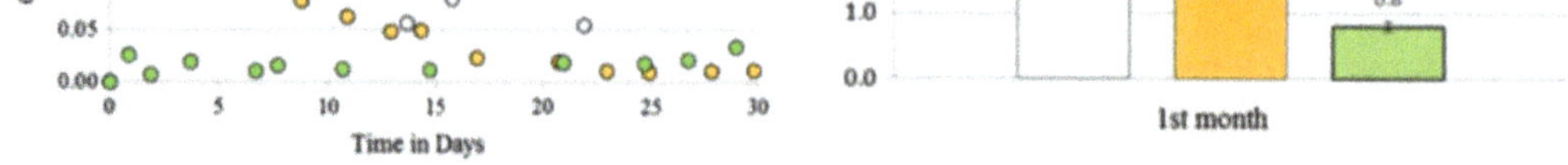

Fig. 10.3 CO_2 production rate in g/h and per kg dry matter (left) and dry matter losses (right) along incubation time for forest residues at a test temperature of 20 °C

CO_2 production decreased steadily, whereas at higher moisture contents, there was still a high CO_2 production rate, which remained static at 0.1 g (h^{-1} kg^{-1}).

The present study shows the moisture content is a key factor of the wood degradation process. The decomposition of sugars and other wood components occur predominantly in the thin liquid films on the wood's surface. Optimal microbe growth takes place at a moisture content between 30 and 60%. A minimum of 20% is crucial for the growth of different fungi [2]. As we could show, drying the biomass has a great effect on the storage stability (low dry matter loss, low risk of self-ignition) but humidification has to be prevented by either a semi-permeable fleece or by storing the biomass in a hall.

10.3.2 *Variation of Particle Sizes*

In order to investigate the influence of particle size distribution on the microbial activity, varying particle sizes were prepared. Figure 10.4 shows the correlation between the microbial activity and the fine-material content over a period of 30 days. The activity of fine and medium woodchips was very similar. On the contrary, the coarse material showed the highest activity and dry matter loss with 4.9%. In this test run, the CO_2 production rate showed strong fluctuations, which probably is due to changes in temperature (22 ± 5 °C). Further experiments have to be conducted including single particle size fractions.

10.3.3 *Variation of the pH*

Figure 10.5 shows the CO_2 production rate and dry matter losses over time. Adding 5% (w/w) $CaCO_3$ increased the pH value to 8.5 and 10% (w/w) $CaCO_3$ increased pH to 9.0.

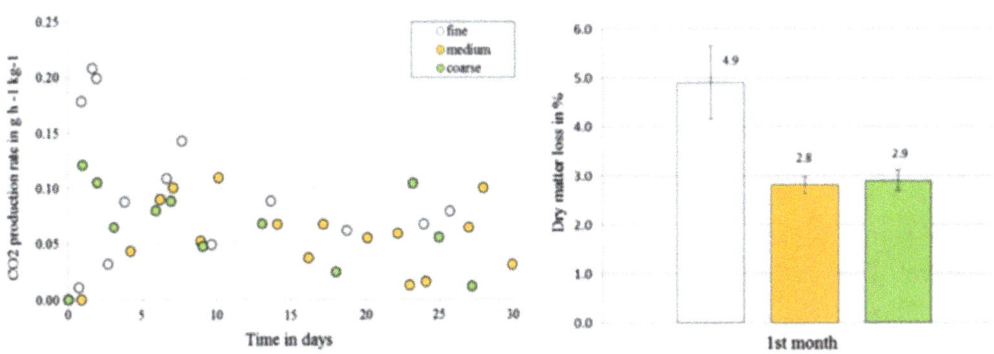

Fig. 10.4 CO_2 production rate in g/h per kg of dry matter (left) along incubation time for fine, medium and coarse forest residues and the dry matter losses (right)

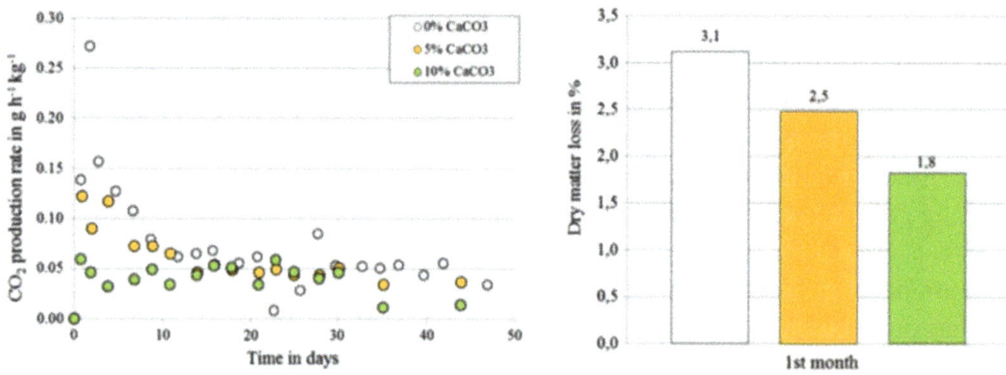

Fig. 10.5 CO_2 production rate in g/h per kg dry matter and dry matter losses along incubation time for forest residues with 0, 5 and 10% $CaCO_3$ (temperature 20 °C, moisture content 47%)

A clear difference in the microbial activity was seen within the first two weeks of incubation. Adding $CaCO_3$ to the woodchips led to a suppression of fungal growth and activity. The dry-matter loss (in w/w) in the first month amounted to 3.1 for 0% $CaCO_3$, 2.5 for 5% $CaCO_3$ and 1.8 for 10% $CaCO_3$. However, no repetition of the experiments have been done so far. Further long-term tests and different additives as well as concentrations have to be investigated. However, a rise in pH did inhibit most of the mold fungi, leading to minimal wood degradation. Thus, adding alkaline substances to the woodchips during storage might suppress the initial temperature increase and lower self-ignition risk. Our previous investigation showed a direct link of microbial activity and a temperature rise within the first week of storage. This assumption has to be investigated in a further project. Additives such as chalk and dolomite not only have the potential to inhibit fungal wood decay but also are used as agents for the combustion of agrarian residues, such as straw, leading to a higher ash melting point.

10.4 Conclusion and Outlook

Respirometric tests on spruce forest residues were conducted under different moisture contents (15–50%), particle sizes (fine, medium, coarse) and pH values (addition of 5 and 10% of $CaCO_3$). The moisture content played a major role during the degradation process. Drying the woodchips to less than 30% reduced the microbial activity and dry-matter loss significantly. At below 20%, no activity could be verified. Increasing the pH to a minimum value of 9 inhibited fungal growth significantly, offering a good alternative for conserving biomass. The influence of the particle size distribution was less distinct as expected but further experiments are crucial for gaining more insight into this.

It can be concluded that respirometric tests provide an accurate way to determine the wood degradation rate, showing a high potential to investigate the wood degradation process. Further improvements have to be completed especially if different test temperatures will be investigated. It is possible to determine the microbial activity over time, with the highest activity occurring within the first few days of storage. This initial activity is responsible for the very high temperatures inside the woodchip piles, which could reach up to 80 °C. Further experiments will be conducted with different test materials and under varying test conditions. In parallel, soil-block decay tests will be conducted and the wood degradation will be investigated by chemical and thermal gravimetric analyses. Additionally, the degraded biomass will be tested in a small scale gasification plant do investigate the influence on the gasification and combustion process and the obtained product gas.

Acknowledgements This work was supported by the government of Tyrol and the Tiroler Standortagentur of the research program, Tiroler Kooperationsförderung. Thanks are due to the partners of this project, namely MCI-The Entrepreneurial School®, SYNECO-tec GmbH, Syneco Consulting, Südtiroler Energieverband and Bioenergie Tirol Nahwärme GmbH, with all district heating plants being part of the research.

References

1. M. Noll, R. Jirjis, Microbial communities in large-scale wood piles and their effects on wood quality and the environment. Appl. Microbiol. Biotechnol. **95**(3), 551–563 (2012)
2. A.T. Martínez, M. Speranza, F.J. Ruiz-Dueñas, P. Ferreira, S. Camarero, F. Guillén, Biodegradation of lignocellulosics: microbial, chemical and enzymatic aspects of the fungal attack of lignin. Int. Microbiol. **8**, 195–204 (2005)
3. Wd Boer, L.B. Folman, R.C. Summerbell, L. Boddy, Living in a fungal world: impact of fungi on soil bacterial niche development. FEMS Microbiol. Rev. **29**(4), 795–811 (2005)
4. V. Scholz, C. Idler, W. Daries, J. Egert, Lagerung von Feldholzhackgut. Agrartechnische Forschung **4**, 100–113 (2005)
5. T. Thörnqvist, Drying and storage of forest residues for energy production. Biomass **7**(2), 125–134 (1985)
6. M. Barontini, A. Scarfone, R. Spinelli, F. Gallucci, E. Santangelo, A. Acampora et al., Storage dynamics and fuel quality of poplar chips. Biomass Bioenergy **62**, 17–25 (2014)
7. C. Whittaker, W. Macalpine, N.E. Yates, I. Shield, Dry matter losses and methane emissions during wood chip storage: the impact on full life cycle greenhouse gas savings of short rotation coppice willow for heat. Bioenergy Res. (2016)
8. X. He, A.K. Lau, S. Sokhansanj, C. Jim Lim, X.T. Bi, S. Melin, Dry matter losses in combination with gaseous emissions during the storage of forest residues. Fuel **95**, 662–664 (2012)
9. R. Buggeln, Outside storage of wood chips. Biocycle **40**, 32–34 (1999)
10. R. Pecenka, H. Lenz, C. Idler, W. Daries, D. Ehlert, Development of bio-physical properties during storage of poplar chips from 15 ha test fields. Biomass Bioenergy **65**, 13–19 (2014)
11. S.M. Kazemi, D.J. Dickinson, R.J. Murphy, Effects of initial moisture content on wood decay at different levels of gaseous oxygen concentrations. J. Agric. Sci. Technol. **3**, 293–304 (2001)
12. F. Ferrero, M. Malow, M. Noll, Temperature and gas evolution during large scale outside storage of wood chips. Eur. J. Wood Prod. **69**(4), 587–595 (2011)
13. M.A. Brand, M. Bolzon de, I. Graciela, W.F. Quirino, J.O. Brito, Storage as a tool to improve wood fuel quality. Biomass Bioenergy **35**(7), 2581–2588 (2011)

14. T. Filbakk, G. Skjevrak, O. Høibø, J. Dibdiakova, R. Jirjis, The influence of storage and drying methods for Scots pine raw material on mechanical pellet properties and production parameters. Fuel Process. Technol. **92**(5), 871–878 (2011)
15. R. Pecenka, H. Lenz, C. Idler, W. Daries, D. Ehlert, Efficient Harvest and Storage of Field Wood for Profitable Production of Bioenergy from Short Rotation Coppice, in *21st European Biomass Conference and Exhibition*, Kopenhagen, Denmark 2013, pp. 120–124
16. R. Jirjis, Storage and drying of wood fuel. Biomass Bioenergy **9**(1–5), 181–190 (1995)
17. Determination of the ultimate aerobic biodegradability of plastic materials in an aqueous medium—Method by analysis of evolved carbon dioxide (ISO 14852:1999): Bureau of Indian Standards, 1999
18. Austrian Standards. ÖNORM EN 15149-1 - Feste Biobrennstoffe - Bestimmung der Partikelgrößenverteilung - Teil 1: Rüttelsiebverfahren mit Sieb-Lochgrößen von 1 mm und darüber (ÖNORM EN 15149-1), 2010

Chapter 11
Impacts of Electrification Under the Perspective of the Multi-Tier-Framework in Southern Tanzania

Annika Groth

Abstract Off-grid areas in many African countries do not necessarily lack access to electricity. In the last decade, energy technologies based on solar power achieved higher penetration rates, also in rural areas of Sub-Saharan Africa. Mini-grid technologies are expected to play a key role in expanding the access to electricity. However, grid extension is still the preferred technology to enhance electrification rates. Taking into account the Multi-Tier-Framework (MTF) by the World Bank, electricity access is no longer a binary metric but a multi-dimensional phenomena. Reliability is one of the criteria considered in the new framework. This study strives to reflect enhanced reliability through an interconnected mini-grid system by comparing the effects of power outages on households in the Southern Tanzanian Region. The focus of this paper is the daily mean lighting hours consumed per household in both a mini-grid-electrified area and none mini-grid electrified areas. Lighting is one of the most important intermediary outcomes of electricity through which households can benefit in many fields. As has been expected, lighting hours consumed by households in mini-grid-connected areas are affected by power outages but are still significantly higher than in not yet grid-connected villages. The analysis underlines the importance of interconnected systems supporting the reliability of electricity access, which is also crucial for productive uses. Additionally, fertile ground for further research is identified. Propensity Score Matching Method is recommended to identify treatment and control group to further study the impacts of interconnected mini-grid electrification.

Keywords Interconnected energy systems · Electricity access · Multi-tier framework · Reliability · Socio-economic impacts · Sub-Saharan Africa

A. Groth (✉)
Department Energy and Environmental Management, Interdisciplinary Institute for Environmental-, Social- and Human Studies, Europa Universität Flensburg, Flensburg, Germany
e-mail: annika.groth@uni-flensburg.de

M. Mpholo et al. (eds.), *Africa-EU Renewable Energy Research and Innovation Symposium 2018 (RERIS 2018)*, Springer Proceedings in Energy,
https://doi.org/10.1007/978-3-319-93438-9_11

11.1 Introduction

The relationship between (rural) electrification and socio-economic impacts has been studied widely on macro-economic level. However, up to now, there is no clear consensus regarding the causal direction of this relationship [1]. But the relevance for electrification as one of the drivers for achieving the sustainable development goals is not questioned. Also on micro economic level, there is some evidence that electrification improves living conditions in developing countries [2].

According to the binary definition of "having a connection to electricity or not", the African continent is with 587 million Africans (excluding North Africa) out of more than 900 million people or with 63% of them not having access to electricity in 2014 far away from reaching the UN's development target of universal access by 2030 [3]. Tanzania, which is in the focus of this study, still belongs to one of the 20 least electrified countries in the world and most recent data from 2014 indicates that only approximately 16% of its population is electrified [3].

However, as acknowledged in the recently developed United Nations Sustainable Energy for All Global Tracking Framework the binary definition of electricity access is too narrow to describe the complexity of it. Energy access-which contains the access to electricity—should be adequate, available when needed, reliable, affordable, legal, convenient, healthy and safe for all required energy applications [4].

This study considers the reliability of electricity by taking into account the duration and frequency of power outages and its impacts on lighting hours of households. On the other hand, the analysis studies households from not yet grid connected areas to reflect their "pre-grid electrification status" allowing access to basic electricity services. With worldwide falling prices for solar power based technologies, ex ante grid electricity based on alternatives to for example diesel generators becomes also more accessible for poorer households in rural areas of developing countries. In rural Sub-Saharan regions many of recently electrified households still use electricity mainly for lighting purposes [5, 6]. Lighting is seen as an intermediary outcome of electrification with the potential to improve final outcomes in the field of health, education and income in the long run.

In the next section, the article reviews research done in the field before it reflects the methodology applied and discusses the results. Finally, it concludes and gives an outlook on further research.

11.2 Background

Tanzania belongs to one of the African countries with a stable economic growth rate of 7% annually in the last decade [7]. The agricultural sector is the backbone of the economy employing more than two thirds of the population [8] which amounts to 55.6 million people in 2016 [7]. The country is still one of the poorest countries in the world, reflected in position number 151 out of 188 countries in the Human

Development Index (HDI) [9] and the Multi-Dimensional Poverty Index (MPI), which defines 66.4% of the Tanzanian population as multi-dimensionally poor in terms of education, health and standard of living [9].

This is also reflected in the official electrification rate, which defines that only 16% of the population is electrified. Per capita electric power consumption amounts to approximately 99 kWh [7]. Installed power generation capacity is low with only 1564 MW [10], whereby approximately 10% of it is attributed to mainly fossil fuel powered mini-grid systems [11].

However, the Tanzanian energy sector is frequently affected by power generation outages, which can be attributed to chronic underinvestment and weak technical as well as financial performance [12] but also to climatic conditions due to its high dependence on hydro power (more than 30% of total generation capacity) [13]. In 2014, approximately 18% of electric power transmission and distribution has been lost [7].

To address these constraints, the Tanzanian government put ambitious reforms into place which include a higher participation of independent power producers (IPP) and small power producers (SPPs) in the power generation sector. The Mwenga Hydro Power Project (Mwenga in the following), which is in the focus of this study, is a 4 MW hydro power based interconnected mini-grid system and falls under the umbrella of a "special regulatory framework with simplified procedures and standardized contracts" [10]. The majority of its power is sold to the main grid (the national utility called Tanesco). The rest of the power generated is distributed within the mini-grid system which encompasses the local tea industry and surrounding rural villages. The shares of what is distributed within the mini-grid or sold to the national grid fluctuates depending on season.

Commonly, research done in the field of impact evaluation of (rural) electrification is on different levels: macro and/or micro level. Irrespective of research level, it focusses mainly on off-grid or grid electrification and rarely studies the effects of interconnected electrification projects. Additionally, the impacts of power outages on (intermediary) impact indicators—such as lighting hours—have been studied less, especially in the Sub-Saharan context.

11.3 Methodology

In 2015, 327 households and enterprises were interviewed in mini-grid connected or not yet mini-grid connected areas in the Mufindi Region in Iringa located in the Southern Tanzanian Highlands [14]. By that time the Mwenga Project already operated for three years. The surveys contained more than 70 detailed questions on socio-economic conditions and energy use. For the purpose of this study, questions related to households' sources of energy use and daily average usage in hours were analyzed. An overview on these questions can be found in the Annex in Tables 11.4 and 11.5.

Household and enterprise selection was based on simple random selection. However, the selection of villages was not randomized because the author wanted to ensure that the villages share most of their background characteristics to enhance comparability between households from the mini-grid connected and not yet mini-grid electrified villages. For that purpose, the village selection procedure considered accessibility of villages, the existence of complementary infrastructure and context characteristics such as topography, distance to bigger cities and towns, educational services, health services (regular) markets in the village, (formal) financial services, mobile phone network, main income sources and presence of other development projects. To get a comprehensive overview on the background characteristics information from different sources was collected and combined. These included local informants like village leaders or representatives from the Mwenga Project and secondary information like official reports [15, 16] and other studies [17].

The present study limits its analysis on 40 mini-grid connected households relying solely on electricity for lighting purposes and 68 households from not yet grid electrified villages.

The concept of reliability of electricity is based on the definition of the World Bank within the Multi-Tier-Framework [4]. In accordance with this concept a non-reliable electricity access is understood here as the time electricity distribution of the mini-grid system is interrupted. The higher the frequency and time of interruptions the more unreliable the supply of electricity becomes.

Data on mini-grid power outages is based on information from project representatives [18]. Power outages refer to the time mini-grid distribution of electricity is interrupted, thus no power is delivered to the end consumers (to the mini-grid connected villagers and to the main grid), irrespective on mini-grid running mode—interconnected or island mode. Thus, power outages from the main grid are not reflected totally because the interconnected system is able to disconnect from the main grid and to switch on isolate mode to further distribute to the villages. Especially planned power outages by the main grid are therefore not reflected here because the system is prepared to switch to isolate mode. However, the data describes unplanned power outages and the time needed to switch the operation to an isolate operation of the mini-grid. Power outages due to occurrences within the Mwenga system are reflected totally. For the purpose of this analysis mini-grid power outages attributable to Tanesco or Mwenga are calculated in average hours per day between 7 p.m. and 6 a.m. on a yearly basis first (see Table 11.1). It is assumed that within that time frame household's lighting hours might be impacted by power outages. At this stage of research, no impacts of seasons or other parameters affecting mini-grid power distribution are reflected.

To better reflect seasonal fluctuations and extraordinary events affecting power generation and distribution, average power outages per day between 7 p.m. and 6 a.m. are also displayed on monthly basis (see Fig. 11.1). The estimations on yearly and monthly basis assume that power outages take place every day- an assumption that might be too strong to reflect reality, but is needed to study impacts on daily lighting usage of households.

To address these constraints, the study further includes average outage frequency per month between 7 p.m. and 6 a.m. (see Table 11.3), which is calculated on yearly basis. Additionally, this study displays average daily distribution in kWh within the mini-grid system between 7 p.m. and 6 a.m. to illustrate the effect of power outages on electricity distribution.

Table 11.1 Average power outage duration in hours per day from 7 p.m. to 6 a.m. in 2015 and 2016

	2015 (h)	2016 (h)
Mwenga	0.51	0.19
Tanesco	0.53	0.25
Both combined	1.04	0.44

Source Own elaboration based on [18]

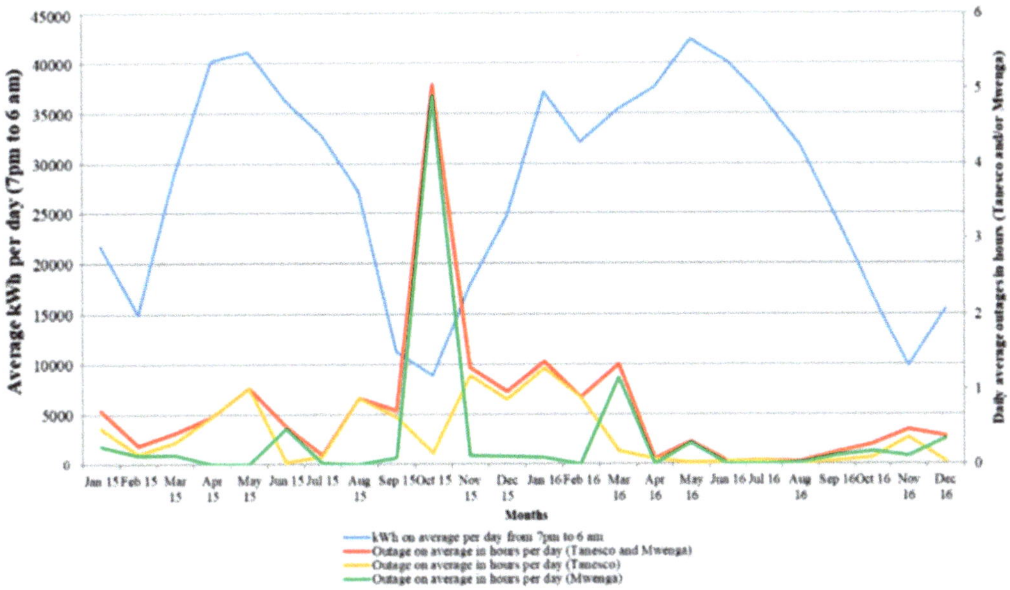

Fig. 11.1 Daily mean power outages in hours and distributed kWh (7 p.m. to 6 a.m.). *Source* Own elaboration based on [18]

Average lighting hours per day are based on estimations of household heads from the non-connected and mini-grid connected villages. Lighting is a direct outcome of electrification because its usage usually starts immediately after electrification when infrastructure and lighting devices are installed. Data has been collected by the end of 2015. For that reason, it was assumed that average lighting usage remained the same for 2016. Lighting hours are based on the daily usage of the most frequent lighting devices or appliances, such as different electric bulb types and wick or gas powered lamps. Due to data constraints mobile torches, such as mobile phone flashlights, and candles have been excluded from the analysis.

11.4 Results and Discussion

As can be seen in Table 11.1, the mean duration of power outages between 7 p.m. and 6 a.m. affecting mini-grid distribution attributed to Mwenga are less compared to those related to Tanesco in 2015 and 2016.

In a worst case scenario, where both power outages would have taken place on the same day but not necessarily at the same time, mini-grid distribution of electricity would have been interrupted by approximately 1 h per day in 2015 and 0.4 h per day in 2016 on average.

With approximately 32.95 mean hours of lighting per day households from mini-grid connected villages consume significantly more lighting hours per day than households from not yet—grid connected areas with 23.94 mean hours per day (see Table 11.2).

On average, mini-grid connected households reported to own 5.3 electric lighting appliances. If their usage in hours is assumed to be equally distributed, this would lead to a usage of each device for approximately 6.2 h per day. For those mini-grid connected households solely relying on electricity from the grid, power outages impact their lighting consumption.

In the worst case scenario, assuming that all lighting devices are running when both power outages take place, this would lead to approximately 31.9 lighting hours per day on average in 2015 and to 32.5 mean lighting hours in 2016.

When distinguished by source of outage, power outages that are attributable to the main grid would have led to 32.42 mean lighting hours in 2015 and 32.7 average lighting hours in 2016.

Conversely, power outages from the Mwenga system would have led to slightly higher mean lighting hours with 32.44 average lighting hours in 2015 and to 32.77 mean lighting hours in 2016. However, the slight differences in average lighting hours are not significant when distinguished by source of outage.

Due to data constraints it is not possible to refine the analysis in terms of a better reflection of real lighting hours diminished by power outages expressed in hours per lighting device. This becomes especially clear, when considering the fact that power outages are not taking place every day. Average outage frequency per month between 7 p.m. and 6 a.m. amounts to 13 in 2015 and to 11 in 2016, which leads to higher average duration per power outage (2.4 h in 2015 and 1.4 h in 2016) and reflects that end users are not affected by daily power outages in a month (see Table 11.3).

However, a comparison with World Bank data on power outages in the national grid in a typical month from 2013 reveals that the interconnected mini-grid system seems to distribute power in a more reliable manner. Tanzanian enterprises reported to be affected by only approximately 9 power outages per month with an average duration of 6.3 h each [19].[1]

The blue line shown in Fig. 11.1 indicates average daily distribution in kWh within the mini-grid system and reflects seasonal and/or extraordinary events and power outages: In dry seasons, from end of June until the end of December, the production from the Mwenga Project and its mini-grid distribution is substantially reduced. In this period, the share of electricity distributed to the village customers (via mini-grid) amounts to approximately 20%, whereas 80% is distributed to the

[1]But in this context, it is important to consider that World Bank's data is based on enterprises surveys and might not reflect power outages within the time frame between 7 p.m. and 6 a.m.

Table 11.2 Households average lighting hours per day

	No outages considered	Both outages combined 2015	Both outages combined 2016	Tanesco outage 2015	Mwenga outage 2015	Tanesco outage 2016	Mwenga outage 2016
Mini-grid connected households	32.95	31.9	32.5	32.42	32.44	32.7	32.77
Not yet grid-connected households	23.94 h***	NA	NA	NA	NA	NA	NA

***, **, * indicate 1%, 5% and 10% levels of level of significance respectively

Source Own elaboration based on [14, 18]

Table 11.3 Monthly mini-grid average outage frequency and duration from 7 p.m. to 6 a.m. in 2015 and 2016

	2015	2016
Average frequency	13	11
Average duration	2.4 h	1.4 h

Source Own elaboration based on [18]

main grid. This is mainly attributed to the irrigation practices of the anchor customers from tea production companies during that season.

In wet seasons, from end of December until the end of June, 90% of the electricity produced is distributed to the main grid and the remaining 10% is distributed to the local villages within the mini-grid [18]. Extraordinary events can also be studied in Fig. 11.1. The red, green and yellow lines indicate the daily mean power outage in hours. The green line shows outages from the Mwenga system. Between October and November 2015 comprehensive maintenance work on the system was undertaken which explains the outliers displayed here.

The flexibility in terms of distribution according to seasonal fluctuations illustrates the advantages of an interconnected system which is able to adapt its distribution to seasonal or extraordinary events. A counterbalancing effect can be identified when distribution is maintained in case of failure in one of the interconnected systems. Thereby, reliability of electricity access can be enhanced.

11.5 Conclusion and Recommendations

Interconnection of mini-grid system and main grid can be beneficial for households. This can be achieved through enhanced reliability of electricity by adapting the distribution to seasonal and/or extraordinary events and power outages, e.g. by switching to island mode in case of failure of the main grid. Lighting hours of households are significantly higher in mini-grid connected villages compared to not yet grid-connected areas. However, their lighting hour consumption is limited by frequent power outages which can be counterbalanced by the interconnection of the system. To further study impacts of power outages on the intermediary outcome of lighting, data on power outages from the main grid and households from main grid connected areas could be collected. The application of more profound statistical methods could allow for more robust results, e.g. a propensity score matching analysis could help to identify counterfactual and research groups. Furthermore, more socio-economic indicators could be included in the analysis as well as a study on the effects of power outages and interconnected systems on small and medium enterprises. The inclusion of lumen hours could additionally give a better reflection on the quality of lighting.

Annex

See Tables 11.4 and 11.5.

Table 11.4 Household survey on energy sources and usage

	Q1. Which of the following energy sources does this household use? Multiple entries are possible. READ ENERGY SOURCE IF ENERGY SOURCE IS USED, ASK Q2-Q5 FOR THIS ENERGY SOURCE BEFORE ASKING Q1 FOR NEXT ENERGY SOURCE												Q2. For which of the following purposes do you use __? Multiple entries are possible. FILL ENERGY SOURCE TAPPED AS INDICATED IN Q1 THEN READ PURPOSE	Q3. Last week or month, roughly how much did this household spend on __? FILL ENERGY SOURCE TAPPED AS INDICATED IN Q1.	Q4. Last week or month, roughly how much did this household use of __? FILL ENERGY SOURCE TAPPED AS INDICATED IN Q1.	Q5. Roughly how many minutes per day or per week does this household spend acquiring __? FILL ENERGY SOURCE TAPPED AS INDICATED IN Q1.
ENERGY SOURCE *HBW = Home based working	1	2	3	4	5	6	7	8	9	10	11	12				
	Lighting	Cooking	TV	Washing	Radio	Computer	Charging	Water Supply	Timber	Agro –processing	Other (HBW*)	Other				
Dry cell batteries																
Hand-crafted																
Car or other rechargeable battery																
Gas (LPG / LNG)																
Diesel (Non-vehicle-operation-use)																
Petrol (Non-vehicle-operation-use)																
Diesel (vehicle-operation-use)																
Petrol (vehicle-operation-use)																
Paraffin / Kerosene																
Candles																
Biogas																
Charcoal / briquettes																
Crop residue (bought)																
Crop residue (collected)																
Firewood (bought)																
Firewood (collected)																
PV-System (Solar)																
Electricity from mini-grid																
Other																

*Home based work

Source Own elaboration based on [14, 15, 20]

Table 11.5 Household survey on lighting devices and daily average usage

Q1. How many of the following lighting devices does this household use?		Q2. What is the mean number of hours you use this _____ per day? FILL ONLY WITH LIGHTING DEVICE USED AS INDICATED IN Q1
Lighting device	Quantity	Total hours used per day
Energy saver		
Incandescent bulb (<50 W)		
Incandescent bulb ($\geq$ 50 W)		
Fluorescent tube		
Solar lamp		
LED lamp (mobile)		
LED lamp (torches)		
Pressurized lantern		
Wick lamp (paraffin/kerosene)		
Gas lamp		
Candle (per week)		
Other (specify):		

Source Own elaboration based on [14, 15, 20]

References

1. A. Omri, An international literature survey on energy-economic growth nexus. Evidence from country-specific studies. Renew. Sustain. Energy Rev. **38**(Supplement C), S. 951–959 (2014). https://doi.org/10.1016/j.rser.2014.07.084
2. Independent Evaluation Group, The Welfare Impact of Rural Electrification: The World Bank, 2008. https://siteresources.worldbank.org/EXTRURELECT/Resources/full_doc.pdf. Accessed on 11.01.2018
3. World Bank, Sustainable Energy for All 2017—Progress toward Sustainable Energy, 2017. https://www.seforall.org/sites/default/files/eegp17-01_gtf_full_report_final_for_web_posting_0402.pdf. Accessed on 11.01.2018
4. M. Bhatia, N. Angelou, Beyond connections: energy access redefined. ESMAP Technical Report, 008/15. World Bank, Washington, DC. © World Bank. https://openknowledge.worldbank.org/handle/10986/24368, License: CC BY 3.0 IGO. Accessed on 11.01.2018
5. G. Bensch, M. Kreibaum, T. Mbegalo, J. Peters, N. Wagner, The status of energy access in three regions of Tanzania: baseline report for an urban grid upgrading and rural extension project. RWI Materialien, No. 111 (2016). ISBN 978-3-86788-781-6
6. L. Lenz, A. Munyehirwe, J. Peters, M. Sievert, Does large-scale infrastructure investment alleviate poverty? Impacts of Rwanda's electricity access roll-out program. World Dev. **89**, S. 88–110 (2017). https://doi.org/10.1016/j.worlddev.2016.08.003
7. The World Bank, World Development Indicators, 2017. http://databank.worldbank.org/data/reports.aspx?source=world-development-indicators. Accessed on 11.01.2018
8. Central Intelligence Agency, Tanzania. In The world Factbook (2017). Retrieved from https://www.cia.gov/library/publications/the-world-factbook/geos/tz.html. Accessed on 11.01.2018
9. United Nations, Human development report, Human development for everyone (United Nations Publications, New York 2016). http://hdr.undp.org/sites/default/files/2016_human_development_report.pdf. Accessed on 11.01.2018

10. African Development Bank Group, Renewable Energy in Africa-Tanzania Country Profile (2015). https://www.afdb.org/fileadmin/uploads/afdb/Documents/Generic-Documents/Renewable_Energy_in_Africa_-_Tanzania.pdf. Accessed on 11.01.2018
11. L. Odarno, E. Sawe, M. Swai, M.J.J. Katyega, A.C. Lee, Accelerating mini-grid deployment in Sub-Saharan Africa: lessons from Tanzania. TaTEDO, World Resources Institute (2017). https://www.wri.org/sites/default/files/accelerating-mini-grid-deployment-sub-saharan-africa_1.pdf. Accessed on 11.01.2018
12. USAID, Investment Brief for the Electricity Sector in Tanzania, Washington D.C. (2015). https://www.usaid.gov/sites/default/files/documents/1860/Tanzania%20_IG_2015_05_03.pdf. Accessed on 11.01.2018
13. The United Republic of Tanzania Ministry of Energy and Minerals, Tanzania's SE4ALL Investment Prospectus, Dar Es Salaam (2015). https://www.seforall.org/sites/default/files/TANZANIA-INVESTMENT-PROSPECTUS.pdf. Accessed on 11.01.2018
14. A. Groth, Comparison of (pre-) electrification statuses based on a case study in Tanzania, in *Conference Paper 0640-1, 11th Conference on Sustainable Development of Energy, Water and Environment Systems—SDEWES Conference*, Lisbon (2016)
15. The United Republic of Tanzania, Basic Demographic and Socio-Economic Profile, 2012 Population and Housing Census, Detailed Statistical Tables, National Bureau of Statistics, Ministry of Finance, Dar es Salaam (2014). http://www.nbs.go.tz/nbstz/index.php/english/statistics-by-subject/population-and-housing-census/249-2012-phc-tanzania-basic-demographic-and-socio-economic-profile. Accessed on 11.01.2018
16. M. Gratwicke, Rift valley energy—small hydro development in Tanzania, ESMAP Knowledge Exchange Forum, The Hague, 13 November 2013
17. Tanzanian Tea Research Institute, Mwenga Hydro Power Project Baseline Study, Dar Es Salaam (2009)
18. D. Protas, Rift valley energy, June 2017, January 2018
19. The World Bank, http://www.enterprisesurveys.org/data/exploreeconomies/2013/tanzania#infrastructure. Accessed online 10.01.2018
20. J. Peters, G. Bensch, C.M. Schmidt, Impact monitoring and evaluation of productive electricity use—an implementation guide for project managers, in *Productive Use of Energy (PRODUSE): Measuring Impacts of Electrification on Micro-Enterprises in Sub-Saharan Africa*, ed. by L. Mayer-Tasch, M. Mukherjee, K. Reiche (Eschborn, 2013)

Chapter 12
Prosumers as New Energy Actors

Rafael Leal-Arcas, Feja Lesniewska and Filippos Proedrou

Abstract This chapter analyses the opportunities that prosumers, as new energy actors, bring to achieving energy security goals in the context of the European Union (EU). In energy governance, there is a progressive top–down diffusion of potential, competences, and leverage across the energy value chain from States and corporate actors towards prosumers. Private and public finance should be attracted and directed to infrastructure schemes that will enable a transition from the traditional centralised power network to the decentralised nexus of smart grids. Technology will play a crucial role in facilitating the role of prosumers in the new market in-the-making.

Keywords Prosumers · Energy security · Sustainable energy · Energy democratization · Decentralized energy

12.1 Introduction

The increasing role of new actors in law-making has received attention since the 1990s [1]. Developments in climate change and environmental law in this era have catalysed innovative governance approaches by non-State actors and international organisations. These developments have created new legal challenges, both public and private, in a global multilevel governance context. New actors are not solely involved in contributing to thematic law and policy agenda setting, developing solutions, and providing oversight capacity; they are also becoming important players in delivering services. Opportunities to deliver services are growing

R. Leal-Arcas (✉)
Queen Mary University of London, London, UK
e-mail: r.leal-arcas@qmul.ac.uk

F. Lesniewska
School of Oriental and African Studies, University of London, London, UK

F. Proedrou
University of South Wales, Pontypridd, UK

M. Mpholo et al. (eds.), *Africa-EU Renewable Energy Research and Innovation Symposium 2018 (RERIS 2018)*, Springer Proceedings in Energy,
https://doi.org/10.1007/978-3-319-93438-9_12

as the global economy reconfigures around advancing information and communications technologies illustrated by the rapidly emerging 'gig' economy.[1]

In this new setting, ample space is created for the emergence of new energy actors, a principal one being prosumers, namely consumers who are also producers of (renewable) energy and who use energy in a smarter and more efficient manner. Energy prosumers is an umbrella term referring to self-generating energy providers, whether households or energy communities. Individuals contribute to the energy supply in their vicinity via their own installed renewable energy capacity, more often than not solar roofing, wind energy, or combined heat and power [2].

This chapter critically analyses the new challenges and opportunities that prosumers bring to achieving energy security goals in the European Union (EU). The EU, along with the United States (US), is a pioneer in engineering a hybrid electricity market model, where traditional power plants will be supplemented by virtual power plants, a plethora of small, individual energy producers and a corresponding new set of mechanisms to cater for the new market. That said, the adoption and customisation of (elements of) this new energy architecture by other countries will hinge upon the degree of its success within European soil. This chapter contributes in two specific ways. First of all, it critically discusses an emerging new actor in the EU's energy security that we refer to as prosumers. Second, it illustrates in broad terms the ways in which this new actor will cooperate with other actors in the EU energy market and contribute to the European Union's energy goals.

In this context, side by side with traditional threats and challenges, new risks, but also opportunities, arise for the ensuring energy security [3]. The energy sector is undergoing a large-scale low-carbon transition. What is underemphasised in this transition is that it involves a major paradigm shift from a supply-driven to a demand-side energy policy. Driven by a mix of geopolitical, economic, climate, and technological considerations, the energy sector is moving towards a new architecture, the principal pillars of which are progressive electrification, a cleaner energy mix, renewable indigenous energy production, increased energy efficiency, and the development of new markets to produce, transmit, and, crucially, manage energy [4]. The key to this overhaul is the slow, but already underway, development of prosumer markets.

[1]'A gig economy is an environment in which temporary positions are common and organizations contract with independent workers for short-term engagements.' (See WhatIs.com (2016) Definition: Gig Economy, updated May 2016, http://whatis.techtarget.com/definition/gig-economy, accessed 12 October 2017.).

12.2 The 'Gig' Economy and New Technologies

The emergence of the 'gig' economy implies the introduction of new actors. The increase in service provisions such as Airbnb, contracting, free-lancing, self-employment, and on-demand Web-based platforms such as Uber are challenging traditionally regulated economic relations. It is not possible to isolate the 'gig' economy per se. Various terms are used, mostly interchangeably, to describe these new economic phenomena including the 'gig' economy, the 'sharing' economy, the 'collaborative' economy, the 'peer-to-peer (P2P)' economy, and the 'access' economy, amongst others. It is arguable whether these notions reflect the same economic model, especially given the plurality and diversity of the activities and the various forms that the 'sharing' scheme may take [5]. Yet, overall the concept itself appears to be simple and has a certain dynamism that fits within the broader context '[t]he advent of the collaborative economy, in combination with artificial intelligence, big data and 3D printing' [6].

To economists, the 'gig' economy represents what is termed as a 'disruptive innovation' [5]. The 'gig' economy offers opportunities for existing and new market players to engage in new forms of economic exchanges. However, it also has negative impacts on current relations among market participants, policymakers, and regulatory authorities. It remains to be seen how this new economy alters the mechanisms of traditional economic schemes.

The growth of the 'gig' economy is intrinsically linked with new technologies. Many innovations depend on access to data at reduced costs. Arguably, there is a chance to maximise economic growth if we have more openly shared data under proper ethical structures, instead of competing data silos.[2] Data become 'most valuable when open and shared'.[3] In the EU, for instance, economic security and growth is associated with the provision of cloud computing. Special Rapporteur Hans Graux claims:

> Allowing easy on-demand access to information technology services, cloud computing can significantly reduce capital expenditure, as cloud users only pay for what they actually use. [… This will foster] innovative business models and services across all industries, generating new advantages for customers and companies alike. […] Small businesses (SMEs) in particular can benefit from the cloud, as they can get access to high-performance IT solutions, which will help them to adapt quickly to new market developments and to innovate and grow their businesses faster [7].

Given this perspective, the cloud has an enormous part to play in decentralised energy provision in the EU energy generation, as it will open up opportunities for new small- and medium-scale actors. To achieve this, Graux envisages a sharing economy that is not held back by regulation and barriers to market access.

---

[2]See the views expressed by Mark Parsons, Secretary-General of Research Data Alliance, in M. Parsons (2017) Letters to the Editor, The Economist, 27 May 2017, at 20.

[3]Ibid.

In the context of a decentralised energy system—a system that places the consumer at the centre of action, empowers the consumer, and therefore democratises the energy system [8]—it is important to talk about smart grids. The term 'smart grids' can be defined in a variety of ways. The following definition is used by the European Regulators' Group for Electricity and Gas (ERGEG), the Council of European Energy Regulators (CEER), and the European Commission in a number of documents:

> [A s]mart grid is an electricity network that can cost efficiently integrate the behaviour and actions of all users connected to it—generators, consumers and those that do both—in order to ensure economically efficient, sustainable power systems with low losses and high levels of quality and security of supply and safety.[4]

Smart grids are integrated systems that include information, technologies, influences from society, managerial supporting arrangements, political limitations as well as legal considerations. Smart metering systems are a stepping stone towards smart grids empowering consumers to actively participate in the energy market. Smart metering systems and smart grids foreshadow the impending 'Internet of Things', and the potential risks associated with the collection of detailed consumption data are likely to increase in the future when combined with data from other sources, such as geo-location data, tracking and profiling on the Internet, video surveillance systems, and radio frequency identification systems.[5]

Across the EU, decentralisation, as well as the development of smart grids and smart metering installation, is occurring at differing rates with the purpose of reducing greenhouse gas emissions. In 2014, approximately 42% of European countries already had a strategic roadmap in place for the implementation of smart grids, while 58% did not [9]. Enabling the necessary regulatory reforms across the spectrum of issues will require innovative approaches to law and regulatory design. The implications of smart grids like the 'gig' economy itself for law, regulation, and policymaking are only beginning to be considered [10]. 'Gig'-based economic activity often raises issues with regard to the application of existing legal

[4]See ERGEG, Position Paper on Smart Grids—An ERGEG Public Consultation Paper, E09-EQS-30-04, 10 December 2009, at 12; ERGEG, Position Paper on Smart Grids—An ERGEG Public Conclusions Paper, E10-EQS-38-05, 10 June 2010; European Commission, Communication from the Commission to the European Parliament, the Council, the European Economic and Social Committee and the Committee of the Regions: Smart Grids: from innovation to deployment, COM (2011) 202 final, 12 April 2011; CEER, CEER Status Review on European Regulatory Approaches Enabling Smart Grids Solutions ('Smart Regulation'), C13-EQS-57-04, 18 February 2014.

[5]Article 29 Data Protection Working Party, Opinion 04/2013 on the Data Protection Impact Assessment Template for Smart Grid and Smart Metering Systems ('DPIA Template') prepared by Expert Group 2 of the Commission's Smart Grid Task Force, 00678/13/EN WP205, 22 April 2013, at 5; Council of Europe, Committee of Ministers, Recommendation CM/Rec(2010)13 of 23 November 2010 of the Council of Europe Committee of Ministers to Member States on the protection of individuals with regard to automatic processing of personal data in the context of profiling, CM/Rec(2010)13, 23 November 2010, at 1.

frameworks and blurs established lines between consumer and provider, employee and self-employed, or the professional and non-professional provision of services.

All of this can result in uncertainty over applicable rules, especially when combined with regulatory fragmentation stemming from divergent regulatory approaches at the national or local level. When it comes to energy, fragmentation of policies, regulations, and cooperation platforms is a constant problem in its utilisation and trade, creating barriers to communication and effective solutions [11]. It remains to be seen whether the impact of such disintegration may slow down the spread of smart grids or, on the contrary, whether it will have the potential to boost it in the future. The European Commission has noted that there is a risk that regulatory grey zones are being exploited to circumvent rules designed to preserve the public interest [12]. In reality, the 'gig' economy is merely adding to a pattern of decentring governance and the emergence of a 'post-regulatory' world [13].

12.3 New Energy Actors

The emergence of new actors on the energy market will influence the feasibility of fulfilling goals related to a more effective energy utilisation. Prosumers have the potential of increasing energy efficiency and securing stable energy supplies for a wider range of consumers, including themselves. New opportunities arise for a new type of economic activity, that of energy aggregators, in what seems to be a much more variable business energy landscape. This role can be fulfilled by incumbent market players as well as by new companies that will focus on encouraging their customers' efficient use of energy and contracting the surplus capacity, which they can sell in a 'flexibility package' to the distributors and utilities. Small storage providers can also emerge in an evolving market that needs back-up capacity and last resort solutions to respond to energy supply and demand variability.

A high premium will be paid for such flexibility services, so the corporate rationale is evidently present. Importantly, there are strong grounds for such economic activity to take place at the community (or even at the district/neighbourhood) level, with co-operatives appearing as a potent form of entrepreneurial type of organisation [14]. The energy market increasingly calls for integrated energy services companies which will optimise both digital technology and electricity distribution by means of trading flexibility services [15]. There are reasons to be optimistic about the affordability of technology advancement for future prosumers: when smartphones came out, they were unaffordable; today, around 80% of phones in the US are smartphones [16]. By analogy, the same effect should happen in energy technology.

There are many ways for citizens, small businesses, and communities to contribute to the energy transition, actively participating in different aspects of the energy market to become true 'energy citizens'. Citizens are no longer resigned to the role of passive consumers, but have the potential to be energy producers, or 'prosumers', particularly through self-generation of renewable energy, energy storage, energy conservation, and participation in demand response [17]. From a legal point of view, prosumers are still considered individuals rather than commercial actors. Their coming together as energy communities will necessitate a more commercial legal status. In general, the future legal status of prosumers is one of the issues that remain unsettled and will have to be determined by the upcoming EU energy regulation.

Crucially, the new energy market creates ample opportunities for individuals and households to become energy traders [18]. Either directly *vis-à-vis* established utilities or indirectly through aggregators, prosumers are empowered to trade the energy they have conserved or produced, thus killing two birds with one stone by facilitating flexibility and network optimisation as well as raising extra revenues for themselves. Indeed, the emphasis of the undertaken energy overhaul lies in distributed energy resources (DER), which enhance local generation and flows into the network [19].

The establishment of a prosumers market is very much a work in progress. Smart grids are the hardware that will allow their full-fledged development, while relevant regulation will be its software. The Commission has taken a solid stance in its Winter Package on several key issues:

- First, it provided for consumers' right to consume the renewable electricity they generate without facing undue restrictions. This means that national jurisdictions that still forbid self-generation will be displaced by EU legislation. On top of this, consumers will be empowered to trade the energy they produce to energy companies, this way becoming active participants in the energy markets and pillars of their consolidated resilience.
- Second, a separate type of energy entity, namely energy communities, is explicitly recognized. Prosumers will be granted the right to group and function in the market collectively.
- Third, more information will be provided regarding energy performance and the sources of district heating and cooling systems. This is a key issue if prosumers and energy communities are to be in practice empowered to improve on their energy performance, including both production and consumption as well as trading. Added to the above, further scrutiny will be paid to improving the quality of information consumers will get. This calls for further refinement of the Guarantees of Origin system with regard to energy resources.

Placing prosumers at the centre of energy markets also requires widespread active demand participation by consumers. This boils down to the 'corporatisation', rationalisation, and economisation of consumer behaviour [20]. Demand response is all about consumers making use of the services digital technology provides to better adjust their energy use to their needs and at the same time adapt their energy usage to the most affordable energy price bands throughout each day. This can grant them significant benefits in terms of energy conservation, efficiency, savings, and extra dividends. Critical information on the state of the grid and running prices enables consumers to turn down, for example, the heating system at peak times to save on energy. This way pressure to the grid is relaxed and the consumer is remunerated. On the other hand, making use of cheap energy (when, for example, solar panels and/or wind turbines generate ample energy) leads to valuable energy savings/surplus for trading [21].

Smart applications and dynamic price contracts are different ways to achieve energy efficiency maximisation. With regard to the former, instructing smart appliances, like the washing machine, to operate at the time of the lowest prices within a day is an energy saver. When it comes to the latter, consumers can take note of their consumption patterns and negotiate corresponding contracts with electricity suppliers. Such contracts should be increasingly on offer and become increasingly more sophisticated and flexible, taking into account that competition is expected to mature and become consolidated. A number of pricing mechanisms can be utilised, such as real-time pricing, time-of-use pricing, critical-time pricing, tariff-of-use pricing, and time-variable pricing to reflect market fundamentals and substitute for traditional methods of dealing with supply-demand disequilibria, such as load-shedding and self-rationing.

Implementing dynamic pricing contracts, however, is more easily said than done. From the perspective of utilities and energy services companies in general, the variability, multi-dimensionality, and heterogeneity of energy use render the 'representative agent' approach deficient in a demand response programme. While a one-size-fits-all programme is hardly the solution, designating appropriate contracts for vastly different (and dynamic) consumer profiles is a daunting task. Efficient contracting is hence practically challenging, especially if one also takes into account the supply side. Suppliers themselves differ in the cost (disutility) they have to bear to draw on flexibility services to match demand each time and are naturally loathe to shoulder extra costs [22].

A crucial parameter that will to a great extent determine the scale of optimisation of smart grids will be the successful incorporation of storage capacity in the system in the form of batteries. This remains commercially challenging, with controversy over the efficiency gains lingering. The use of storage batteries will enable the optimal function of the network, not least since it will reduce peak consumption, system-wide generation costs, losses, and network congestions. At the same time, it will lessen the need for further investments in network expansions [23].

Whether electric vehicles are the optimal means of storage is also debatable. The use of double (or old) batteries, logistical issues pertaining to charging periods and infrastructure costs, as well as the optimisation of the network with vehicles plagued off the grid in rush hours, all remain thorny issues to be sorted out in the near future. On the bright side, the spill-over to one of the most pollutant energy sectors —transportation—holds high promise for scaled-up performance in carbon reductions, an outcome direly needed if designated climate measures fail to stabilise the rise of global temperature in time [24].

The modern grid has been operationally grounded on the worst case dispatch philosophy. Given that the supply side was a priori known, utilities had to balance it with demand which could be in most cases predicted to lie within certain bands. In order to tackle any incidents of supply-demand imbalance (owing either to any supply side failure or to an unpredicted surge in demand), a large reserve capacity was retained. While this added to overall costs and carbon emissions, it provided a shield of protection against power cuts and inevitable load-shedding [25].

Following the same principles and rationale hardly makes any sense for smart grids, whose very function is based on the stochastic and dynamic nature of both supply and demand. An increased degree of intermittent renewable energy, unreliability of storage, micro-grids, variability in consumer choice, and the function of smart appliances all increase uncertainty in both supply and demand of electric power. This variability and unpredictability can be mediated and tackled by a number of tools, such as sensors, smart meters, and a wide range of demand response mechanisms that provide accurate information on the state of the power system and the supply-demand equilibrium as well as more refined means of control of energy use [26].

Following this rationale, a reconsideration and ensuing redefinition of risks management seems appropriate. What constitutes acceptable risk must certainly be adjusted to the new operating conditions of smart grids and replace the current measures of risk. In this context, it is necessary to pass from only quantities to quantities plus probabilities in supply–demand information analysis. Demand response of individual consumers has to be aggregated into a probabilistic demand curve, analogous to the generation availability curve of intermittent renewable energy [27]. The focus lies constantly on the fluctuations in the *net load,* the difference between total demand (load) and variable generation. A number of attributes are fundamental when considering balancing the load, such as capacity, ramp rate, duration, and lead time for increasing or decreasing supply as appropriate [28].

Cross-border markets and their potential must also be integrated into risk management analysis. The EU's current cross-border electricity trade has worked well in the day-ahead market. A big part of national markets are now coupled, stimulating price competition, improving balancing, and enhancing back-up capacity and hence resilience. Importantly, cross-border markets should increasingly expand to include non-EU Member States that have joined the Energy Community and can also include capacity from neighbouring states outside the Energy Community. While cross-border capacity is definitely a source to tap from, a number of physical barriers such as congestion, lack of transmission capacity, and/or under utilisation have, in several cases, resulted in sub-optimal transmission returns and hub market differentials, which impede, rather than facilitate, cross-border trade. Moreover, while the traditional timeframe for most of electricity trading has been day-ahead trade, moving towards 'real-time', intra-day trade remains a challenge [29].

12.4 Conclusion and Recommendations

This chapter sheds light on the emergence of a new actor, namely the prosumer, in the EU's energy security arena. Sustainable energy is rapidly becoming an EU special brand, like the protection of human rights, in the quest for looking after the environment. Achieving sustainable energy encompasses the following points: decarbonising the economy, democratising access to energy (namely everyone has the right to participate), digitalisation, diversification of energy supply, and disrupting traditional energy cycles. Leadership is shifting from national politics to local politics and, therefore, power is being decentralised. For instance, when there is a natural disaster in a given neighbourhood, citizens do not contact the head of State or government of the nation, but the mayor of the city. A clear example of this trend towards local politics is the Local Governments for Sustainability platform.

Following trends in the EU towards decentralisation and the emergence of a 'gig' economy, the energy sector is currently undergoing a large-scale transition. One of its core aspects is the progressive top–down diffusion of the potential, competences, and leverage from EU institutions, States, and corporate actors across the energy value chain towards prosumers, who need to be at the centre of the energy transition for it to happen democratically in a bottom–up manner. This phenomenon can be conceptualised as energy democratisation, namely moving away from a few energy companies monopolising access to energy towards energy owned mainly by consumers, making consumers of the utmost importance.

All of this is achievable by shifting the current paradigm to one that is more human-centric, by linking projects to people, and more collaborative in how it tackles various obstacles, whether legal or behavioural. Think of the analogy of organic food: it is more expensive, but for many, its benefits outweigh the costs. Moreover, consumers have the power to choose either organic or non-organic. By the same token, many citizens are interested in climate-friendly products even if they are more expensive. This means that we need to look at the whole production process, not just the end product, if we are serious about consumer empowerment. To get there, legislation must remove barriers to participation and protect and promote consumers to enable them to produce, store, sell, and consume their own energy.

While all of the above creates ample potential for facilitating and improving the EU's security of supply as well as fulfilling its climate targets, several caveats exist. These not only are confined within energy security prerogatives, but also extend to the critical management of digital security, which the digitalisation of energy services brings to the fore. So, for consumers to become prosumers and engage in the energy transition, it will be crucial to make the process interesting and simple and to inform them much more, given the current level of energy consumer dissatisfaction. Here is where cities can play a major role at educating citizens on energy transition and climate change mitigation, not least because cities consume three quarters of the world's energy, and because they are smaller entities than countries or regions, so it is easier to get things done. Even more impactful would be to educate companies and policymakers on sustainable development, since there are fewer of them than there are citizens. Doing so will shift the paradigm from a system that is producer-centric to one that will be consumer-centric. This paradigm shift is crucial because development is not possible without energy and sustainable development is not possible without sustainable energy.

In an ever-shifting context, demand management emerges as a key issue. The provision of adequate and precise information to prosumers—so that they can optimise their use of smart grids—as well as the transition to targeted, flexible contracts to adjust to the needs of prosumers need to be embedded in well-articulated broader policy and market regulatory frameworks. Moreover, private and public finance should be effectively attracted and directed to indispensable infrastructure schemes that will enable the transition from the traditional centralised power network to the decentralised nexus of smart grids. And it is well known that where finance flows, action happens. Last but not least, the technologies that will be prioritised in terms of energy generation to back renewable energy generation will play a crucial role in facilitating the role of prosumers in the new market in-the-making. Since renewable energy is becoming more competitive, more green jobs will be created in the future and the trend towards a clean energy revolution is ever closer. This energy transition into renewable energy, in turn, will help both enhance energy security and mitigate climate change. So rather than investing large amounts of money into building liquefied natural gas terminals and gas pipelines, the EU should make a greater effort to invest in renewable energy.

The emerging establishment of prosumer markets is an invaluable development that will enable the transition from supply-driven to demand-side EU energy policy. This cannot but have far-reaching ramifications for the amply politicised and securitised gas trade with Russia as well as for furthering the internal EU market architecture. It is expected that it will decrease flows of energy as well as dependence on Russian gas in the medium term while at the same time acting as a stimulus for further market integration in the energy, climate, and digital economy realms.

Giving civil society a greater voice is imperative for the energy transition to happen. Below are some of the necessary actions:

1. Speeding up action on the ground and localising global agendas;
2. More alliances between countries and donors in the decarbonisation process;
3. Greater collaboration between civil society, governments, and NGOs to include all layers of governance;
4. Bringing together different camps of governments;
5. Scaling up the capacity of local governments;
6. Webbing[6] will be necessary: we need to look at issues and challenges, not sectors; temporal linkages are required, namely using time as an indicator given

[6]By webbing, we are referring to connecting different issues in a broader policy approach rather than approaching them in silos.

its importance in the context of decarbonisation, and there needs to be policy coherence.

Finally, in the future, energy will be consumed near where it is produced. How will this impact international trade (in energy)? Furthermore, the protectionist concept of 'buy local' seems to be going global. This policy is suggested, among other things, to reduce greenhouse gas emissions from transportation, which will benefit climate change. But what implications will it have for international trade? Unless there is more innovation in transportation, there is a chance that this policy will result in less demand for international trade. How can international trade and climate change mitigation work together harmoniously without impeding each other in the context of an emerging decentralised energy system? New actors and modes of governance are changing the traditional global trading system, or at least are contributing to the transformation from inter-State dealings to completely different forms of governance in which non-State actors (including individuals) play a role. The EU has been a social laboratory to test hypotheses of multi-level governance in the past, which are pertinent for the case of energy transition. The above questions are all very relevant to a future research agenda in the broad field of international economic law and governance.

Acknowledgements The financial help from two EU grants is greatly acknowledged: Jean Monnet Chair in EU International Economic Law (project number 575061-EPP-1-2016-1-UK-EPPJMO-CHAIR) and the WiseGRID project (number 731205), funded by Horizon 2020. Bothy grants have been awarded to Professor Dr. Rafael Leal-Arcas.

References

1. A. Clapham, Non-state actors, in *Post-Conflict Peacebuilding: A Lexicon*, ed. by V. Chetail (Oxford University Press, New York, 2009), pp. 200–212; F. Halliday, The romance of non-state actors, in *Non-state Actors in World Politics*, ed. by D. Josselin, W. Wallace (Palgrave, New York, 2001), pp. 21–40; D. Josselin, W. Wallace non-state actors in world politics: a framework, in *Non-State Actors in World Politics*, ed. by D. Josselin, W. Wallace (Palgrave, New York, 2001), pp. 1–20
2. J. Roberts, in *Prosumer Rights: Options for a Legal Framework Post-2020* (ClientEarth, May 2016), http://www.greenpeace.org/eu-unit/Global/eu-unit/reports-briefings/2016/ClientEarth%20Prosumer%20Rights%20-%20options%20for%20a%20legal%20framework%20-%20FINAL%2003062016.pdf
3. R. Leal-Arcas, Mega-regionals and sustainable development: the transatlantic trade and investment partnership and the trans-pacific partnership. Renew. Energy Law Policy Rev. **6** (4), 248–264 (2015)
4. R. Leal-Arcas, J. Wouters (eds.), *Research Handbook on EU Energy Law and Policy* (Edward Elgar, Cheltenham, 2017)
5. V. Hatzopoulos, S. Roma, Caring for sharing? The collaborative economy under EU law. Common Market Law Rev. **54**(1), 81–127 (2017)
6. K. Schwab, The fourth industrial revolution: what it means and how to respond, foreign affairs, 12 December 2015, https://www.foreignaffairs.com/articles/2015-12-12/fourth-industrial-revolution. Accessed 12 Oct 2017

7. H. Graux, Rapporteur to the European Cloud Partnership Steering Board: Establishing a trusted cloud Europe (European Commission, 4 July 2014), https://publications.europa.eu/en/publication-detail/-/publication/b5c80ddb-fa1a-465b-a8f3-3e6c90af4a3b. Accessed 12 Oct 2017, at 8
8. European Commission: 7th citizens' energy forum conclusions, 12–13 March 2015, https://ec.europa.eu/energy/sites/ener/files/documents/2015_03_13_LF_conclusions.pdf. Accessed 12 Oct 2017
9. CEER: CEER status review on European regulatory approaches enabling smart grids solutions ('smart regulation') (2014)
10. D.E. Rauch, D. Schleicher, Like Uber, But for Local Governmental Policy: The Future of Local Regulation of the 'Sharing Economy'. George Mason University Law and Economics Research Paper Series (2015). https://www.law.gmu.edu/assets/files/publications/working_papers/1501.pdf. Accessed 12 Oct 2017; C. Koopman, M. Mitchell, A. Thierer, The sharing economy and consumer protection regulation: the case for policy change. J. Bus. Entrepreneurship Law **8**(2), 529–545 (2015); V. Katz, Regulating the sharing economy. Berkeley Technol. Law J. **30**(4), 1068–1125 (2015)
11. R. Leal-Arcas, A. Filis, The fragmented governance of the global energy economy: a legal-institutional analysis. J. World Energy Law Bus. **6**(4), 348–405 (2013)
12. European Commission: Communication from the commission to the European parliament, the council, the European economic and social committee and the committee of the regions: a European agenda for the collaborative economy, COM(2016) 356 final, 2 June 2016
13. J. Black, Decentring regulation: understanding the role of regulation and self-regulation in a 'post-regulatory' world. Curr. Legal Probl. **54**(1), 103–146 (2001)
14. Y. Cai, T. Huang, E. Bompard, Y. Cao, Y. Li, Self-sustainable community of electricity prosumers in the emerging distribution system. IEEE Trans. Smart Grid **8**(5), 2207–2216 (2016)
15. L. Boscán, R. Poudineh, in *Flexibility-Enabling Contracts in Electricity Markets* (Oxford Energy Comment, 2016), https://www.oxfordenergy.org/wpcms/wp-content/uploads/2016/07/Flexibility-Enabling-Contracts-in-Electricity-Markets.pdf. Accessed 12 Oct 2017
16. Pew Research Center: Mobile fact sheet, 12 January 2017, http://www.pewinternet.org/fact-sheet/mobile/. Accessed 12 Oct 2017
17. J. Roberts, in *Prosumer Rights: Options for a Legal Framework Post-2020*, at 5 (2016)
18. This approach is quite in contrast to the typical top-down way of trading energy. See, for instance, R. Leal-Arcas, How governing international trade in energy can enhance EU energy security. Renew. Energy Law Policy Rev. **6**(3), 202–219 (2015); R. Leal-Arcas, C. Grasso, J. Alemany Rios, Multilateral, regional and bilateral energy trade governance. Renew. Energy Law Policy Rev. **6**(1), 38–87 (2015)
19. European Commission: Proposal for a directive of the European parliament and of the council on the promotion of the use of energy from renewable sources (recast), COM(2016) 767 final, 30 November 2016
20. For further analysis on the link between behavioral sciences and government, see C.R. Sunstein, in *The Ethics of Influence: Government in the Age of Behavioral Science* (Cambridge University Press, New York, 2016)
21. European Commission: Proposal for a directive of the European parliament and of the council on the promotion of the use of energy from renewable sources (recast), 2016, at 10
22. L. Boscán, R. Poudineh in *Flexibility-Enabling Contracts in Electricity Markets* (Oxford Energy Comment, 2016), https://www.oxfordenergy.org/wpcms/wp-content/uploads/2016/07/Flexibility-Enabling-Contracts-in-Electricity-Markets.pdf, at 10
23. C. Eid, R. Hakvoort, M. de Jong, Global Trends in the Political Economy of Smart Grids: A Tailored Perspective on 'Smart' for Grids in Transition. UNU-WIDER Working Paper Series (2016). https://www.wider.unu.edu/sites/default/files/wp2016-22.pdf. Accessed 12 Oct 2017

24. International Energy Agency: World energy outlook 2016: executive summary (2016) https://www.iea.org/publications/freepublications/publication/WorldEnergyOutlook2016ExecutiveSummaryEnglish.pdf. Accessed 12 Oct 2017, at 3 and 5
25. P.P. Varaiya, F.F. Wu, J.W. Bialek, Smart operation of smart grid: risk-limiting dispatch. Proc. IEEE **99**(1), 40–57, at 41–43 (2011)
26. L. Boscán, R. Poudineh, in *Flexibility-Enabling Contracts in Electricity Markets* (Oxford Energy Comment, 2016), https://www.oxfordenergy.org/wpcms/wp-content/uploads/2016/07/Flexibility-Enabling-Contracts-in-Electricity-Markets.pdf, at 7–8
27. P.P. Varaiya, F.F. Wu, J.W. Bialek, Smart operation of smart grid: risk-limiting dispatch. Proc. IEEE **99**(1), 40–57, at 55 (2011)
28. L. Boscán, R. Poudineh, in *Flexibility-Enabling Contracts in Electricity Markets* (Oxford Energy Comment, 2016), https://www.oxfordenergy.org/wpcms/wp-content/uploads/2016/07/Flexibility-Enabling-Contracts-in-Electricity-Markets.pdf, at 9
29. D. Buchan, M. Keay, in *EU Energy Policy—4th Time Lucky?* (Oxford Energy Comment, 2016), https://www.oxfordenergy.org/wpcms/wp-content/uploads/2016/12/EU-energy-policy-4th-time-lucky.pdf, at 6–9

www.ingramcontent.com/pod-product-compliance
Lightning Source LLC
LaVergne TN
LVHW082236200926
848484LV00008B/2362